The vibrations of atoms inside crystals – lattice dynamics – are basic to many fields of study in the solid state and mineral sciences, and lattice dynamics are becoming increasingly important for work on mineral stability. This book provides a self-contained text that introduces the subject from a basic level and then takes the reader, through applications of the theory, to research level.

Simple systems are used for the development of the general principles. More complex systems are then introduced, and later chapters look at thermodynamics, elasticity, phase transitions and quantum effects. Experimental and computational methods are described, and applications of lattice dynamics to specific studies are detailed. Appendices provide supplementary information and derivations for the Ewald method, statistical mechanics of lattice vibrations, Landau theory, scattering theory and correlation functions.

The book is aimed at students and research workers in the earth and solid state sciences who need to incorporate lattice dynamics into their work.

INTRODUCTION TO LATTICE DYNAMICS

CAMBRIDGE TOPICS IN MINERAL PHYSICS AND CHEMISTRY

Editors
Dr Andrew Putnis
Dr Robert C. Liebermann

1 *Ferroelastic and co-elastic crystals*
E. K. H. SALJE

2 *Transmission electron microscopy of minerals and rocks*
ALEX C. MCLAREN

3 *Introduction to the physics of the Earth's interior*
JEAN-PAUL POIRIER

4 *Introduction to lattice dynamics*
MARTIN T. DOVE

INTRODUCTION TO LATTICE DYNAMICS

MARTIN T. DOVE
Department of Earth Sciences
University of Cambridge

Published by the Press Syndicate of the University of Cambridge
The Pitt Building, Trumpington Street, Cambridge CB2 1RP
40 West 20th Street, New York, NY 10011-4211, USA
10 Stamford Road, Oakleigh, Melbourne 3166, Australia

First published 1993

A catalogue record for this book is available from the British Library

Library of Congress cataloguing in publication data available

ISBN 0 521 39293 4 hardback

Transferred to digital printing 2004

TAG

For Kate, Jennifer-Anne and Emma-Clare,
and our parents

Contents

Preface

The subject of lattice dynamics is taught in most undergraduate courses in solid state physics, usually to a very simple level. The theory of lattice dynamics is also central to many aspects of research into the behaviour of solids. In writing this book I have tried to include among the readership both undergraduate and graduate students, and established research workers who find themselves needing to get to grips with the subject.

A large part of the book (Chapters 1–9) is based on lectures I have given to second and third year undergraduates at Cambridge, and is therefore designed to be suitable for teaching lattice dynamics as part of an undergraduate degree course in solid state physics or chemistry. Where I have attempted to make the book more useful for teaching lattice dynamics than many conventional solid state physics textbooks is in using real examples of applications of the theory to materials more complex than simple metals.

I perceive that among research workers there will be two main groups of readers. The first contains those who use lattice dynamics for what I might call *modelling studies*. Calculations of vibrational frequencies provide useful tests of any proposed model interatomic interaction. Given a working microscopic model, lattice dynamics calculations enable the calculation of macroscopic thermodynamic properties. The systems that are tackled are usually more complex than the simple examples used in elementary texts, yet the theoretical methods do not need the sophistication found in more advanced texts. Therefore this book aims to be a half-way house, attempting to keep the theory at a sufficiently low level, but developed in such a way that its application to complex systems is readily understood.

The second group consists of those workers who are concerned with displacive phase transitions, for which the theory of *soft modes* has been so successful that it is now essential that workers have a good grasp of the theory of lattice dynamics. The theory of soft modes requires the anharmonic treatment of the theory, but in many cases this treatment reduces to a modified harmonic theory and therefore remains comprehensible to non-theorists. It seems to me that there is a large gap in the literature for workers in phase transitions between elementary and advanced texts. For example, several texts begin with the second quantisation formalism, providing a real barrier for many. It is hoped that this book will help to open the literature on phase transition theory for those who would otherwise have found it to be too intimidating.

I have attempted to write this book in such a way that it is useful to people with a wide range of backgrounds, but it is impossible not to assume some level of prior knowledge of the reader. I have assumed that the reader will have a knowledge of crystal structures, and of the reciprocal lattice. I have also assumed that the reader understands wave motion and the general wave equation; in particular it is assumed that the concept of the wave vector will present no problems. The mathematical background required for the first five chapters is not very advanced. Matrix methods are introduced into Chapter 6, and Fourier transforms (including convolution) are used from Chapter 9 onwards and in the Appendices. The Kronecker and Dirac delta function representations are used throughout. An appreciation of the role of the Hamiltonian in either classical or quantum mechanics is assumed from Chapter 6 onwards. Chapter 11 requires an elementary understanding of quantum mechanics.

It is an unfortunate fact of life that usually one symbol has two or more distinct meanings. This usually occurs because there just aren't enough symbols to go around, but the problem is often made worse by the use of the same symbol for two quantities that occur in similar situations. For example, Q can stand for either a normal mode coordinate or an order parameter associated with a phase transition, whereas $\mathbf{Q}$ is used for the change in neutron wave vector following scattering by a crystal. Given that this confusion will probably have a permanent status in science, I have adopted the symbols in common usage, rather than invent my own symbols in an attempt to shield the reader from the real world.

It is my view of this book as a stepping stone between elementary theory and research literature that has guided my choice of material, my treatment of this material, and my choice of examples. It is a temptation to an author to attempt to make the reader an expert in every area that is touched on in the book. This is clearly an impossibility, if for no other reason than the constraint on the number of pages! Several of the topics discussed in the individual chapters of this book have themselves been the subject of whole books. Thus a book such as this can only hope to provide an introduction into the different areas of specialisation. The constraints of space have also meant that there are related topics that I have not even attempted to tackle; among these are dielectric properties, electronic properties, imperfect crystals, and disordered systems. Moreover, I have had to restrict the range of examples I have been able to include. Thus I have not been able to consider metals in any detail. I therefore reiterate that my aim in writing this book is to help readers progress from an elementary grasp of lattice dynamics to the stage where they can read and understand current research literature with some intelligence, and I hope that the task of broadening into the missing topics will be less daunting in consequence.

Acknowledgements

This book was written over the period 1990–1992. During that time a number of people helped in a number of ways. Firstly I must thank the Series Editor, Andrew Putnis, for encouraging me to transform my lecture handouts into this book, and Catherine Flack of Cambridge University Press for helping in all the practical aspects of this task. A number of people have helped by reading an earlier draft, and suggesting changes and providing encouragement: Mark Harris (Oxford), Mark Hagen (Keele), Mike Bown (Cambridge), Ian Swainson (Cambridge), Björn Winkler (Saclay), and David Price (London). Of course none of these people must share any responsibility for residual errors or for any features that irritate the reader! However, the help given by all these people has been greatly appreciated. The text for this book was prepared in camera-ready format on my word processor at home, but this would not have been possible without the constant help of Pat Hancock.

My own understanding of the subject of lattice dynamics has been helped by a number of very enjoyable collaborations. It has been a great privilege to work with Alastair Rae (my PhD supervisor), Stuart Pawley, Ruth Lynden-Bell and David Fincham (with whom I did my post-doctoral work), Brian Powell and Mark Hagen (with whom I have done all my inelastic neutron scattering work), and Volker Heine, Ekhard Salje, Andrew Giddy, Björn Winkler, Ian Swainson, Mark Harris, Stefan Tautz, David Palmer and Tina Line (colleagues and students in Cambridge).

To all these people, thank you.

Finally, special thanks are due to my wife, two daughters, and parents. Our eldest daughter, Jennifer-Anne, was less than one year old when I started writing in earnest, and during the period of writing our second daughter Emma-Clare was born. It will be readily appreciated that I could never have completed this book without the considerable encouragement, cooperation and patience of my wife Kate. Thank you!

Physical constants and conversion factors

Fundamental constants

c Velocity of light in vacuo = 2.997925×10^8 m s^{-1}
e Electronic charge = 1.60219×10^{-19} C
h Planck's constant = 6.6262×10^{-34} J s
$\hbar$ Planck's constant = $h/2\pi$ = 1.05459×10^{-34} J s
k_B Boltzmann's constant = 1.3807×10^{-23} J K^{-1}
N_A Avogadro's number = 6.0222×10^{23}
R Gas constant = $N_A \times k_B = 8.314$ J K^{-1}
ε_0 Dielectric permittivity of free space = 8.8542×10^{-12} F m^{-1}
m Mass of the neutron = 1.67492×10^{-27} kg

Conversion factors

Different workers use different units for frequencies, and may often interchange frequencies and energies. Frequencies and temperatures can be converted to energy units by multiplication by h and k_B respectively. Energies are often expressed in units per mole by dividing by N_A. An example serves to explain the use of the conversion table below: a phonon of energy 4 meV has a frequency of 0.968 THz or 32.28 cm^{-1}.

	meV	THz	cm^{-1}	K	kJ mol^{-1}
meV	1	0.241797	8.065467	11.60485	96.484
THz	4.135707	1	33.35640	47.99460	399.0295
cm^{-1}	0.123985	0.029979	1	1.438840	11.96261
K	0.086171	0.020836	0.695005	1	8.31434
kJ mol^{-1}	0.010364	0.002506	0.083594	0.120274	1

1
Some fundamentals

We begin by describing the interatomic forces that cause the atoms to move about. The main interactions that we will use later on are defined, and methods for the determination of specific interactions are discussed. The second part of the chapter is concerned with the behaviour of travelling waves in any crystal.

Indications that dynamics of atoms in a crystal are important: failure of the static lattice approximation

Crystallography is generally concerned with the static properties of crystals, describing features such as the average positions of atoms and the symmetry of a crystal. Solid state physics takes a similar line as far as elementary electronic properties are concerned. We know, however, that atoms actually move around inside the crystal structure, since it is these motions that give the concept of temperature, and the structures revealed by X-ray diffraction or electron microscopy are really averaged over all the motions. The only signature of these motions in the traditional crystallographic sense is the *temperature factor* (otherwise known as the *Debye–Waller factor* (Debye 1914; Waller 1923, 1928) or *displacement amplitude*), although diffuse scattering seen between reciprocal lattice vectors is also a sign of motion (Willis and Pryor 1975). The *static lattice model*, which is only concerned with the average positions of atoms and neglects their motions, can explain a large number of material features, such as chemical properties, material hardness, shapes of crystals, optical properties, Bragg scattering of X-ray, electron and neutron beams, electronic structure and electrical properties, etc. There are, however, a number of properties that cannot be explained by a static model. These include:

- thermal properties, e.g. heat capacity;

- effects of temperature on the lattice, e.g. thermal expansion;
- the existence of phase transitions, including melting;
- transport properties, e.g. thermal conductivity, sound propagation;
- the existence of fluctuations, e.g. the temperature factor;
- certain electrical properties, e.g. superconductivity;
- dielectric phenomena at low frequencies;
- interaction of radiation (e.g. light and thermal neutrons) with matter.

Are the atomic motions that are revealed by these features random, or can we find a good description for the dynamics of the crystal lattice? The answer is that the motions are not random; rather they are determined by the forces that atoms exert on each other. The aim of this book is to show that in fact we have a very good idea of the way atoms move inside a crystal lattice. This is the essence of the subject of *lattice dynamics*.

The classical motions of any atom are simply determined by Newton's law of mechanics: *force* = *mass* × *acceleration*. Formally, if $\mathbf{r}_j(t)$ is the position of atom j at time t, then

$$\frac{\partial^2 \mathbf{r}_j(t)}{\partial t^2} = -\frac{1}{m_j} \nabla \varphi_j\left(\mathbf{r}_j, t\right) \tag{1.1}$$

where m_j is the atomic mass, and $\varphi_j(\mathbf{r}_j, t)$ is the instantaneous potential energy of the atom. Equation (1.1) is our key equation. We therefore need some knowledge of the nature of the atomic forces found in a crystal. The potential energy in equation (1.1) arises from the instantaneous interaction of the atom with all the other atoms in the crystal. We will often assume that this can be written as a sum of separate atom–atom interactions that depend only on the distances between atoms:

$$\varphi_j = \sum_i \varphi_{ij}\left(r_{ij}\right) \tag{1.2}$$

where r_{ij} is the distance between atoms i and j, and $\varphi_{ij}(r_{ij})$ is a specific atom–atom interaction. The sum over i in equation (1.2) gives the interactions with all other atoms in the crystal.

Of course, quantum mechanics rather than classical mechanics determines the motions of atoms. But we will see that the main features of lattice dynamics follow exactly from the classical equation (1.1), whilst quantum effects are primarily revealed in the subsequent thermodynamic properties.

Our aim in this chapter is to set the scene for the rest of the book. In the first part we will consider some elementary ideas associated with interatomic potentials. These are essential if we are to use real examples to illustrate the basic ideas we will develop in the following chapters. In the second part we will present the basic formalism for describing the motions of waves in crystals, which we will build upon in the rest of the book.

Interatomic forces

The variety of interatomic forces

The forces between atoms are all ultimately electrostatic in origin. However, when quantum mechanics is taken into account, the different types of forces can have some very different manifestations.

Direct electrostatic

The electrostatic interactions are long range and well understood, and have in general a very simple mathematical representation. The Coulomb energy follows the well known r^{-1} form, and is strongest in ionic systems such as NaCl. However, there is one complicating feature that can sometimes be neglected but which in other cases plays a crucial role in stabilising the structure. This is the existence of *inductive forces*. Consider, for example, a simple spherical ion on a symmetric site in a lattice, such that at equilibrium all the electric fields at the symmetric site cancel out. If the surrounding neighbours move around, they will generate a residual electric field at the symmetric site; also, if this ion moves off its site it will experience an electric field. This residual field will then polarise our ion under consideration, giving it a temporary dipole moment. This dipole moment will then interact with the charges (and moments if they exist) on the neighbouring ions, adding an extra contribution to the energy of the crystal. In calculations of the dynamics of ionic crystals, this induction energy has often been found to be of considerable importance. Such calculations commonly use the *shell model* (Dick and Overhauser 1958; Cochran 1971; Woods et al. 1960, 1963). In this model, the ions are assumed to comprise the rigid core of the nucleus plus the tightly bound inner electrons, and a loosely bound outer layer, or shell, of the remaining electrons. It is then assumed that the shell and core are held together by a harmonic interaction (that is, the energy is proportional to the square of the distance between the centres), which is the same as saying that the polarisation of the ion is directly proportional to the local electric field. The energy between two ions is then the sum of six interactions: *core*(1)–*core*(2), *shell*(1)–*shell*(2), *core*(1)–*shell*(1),

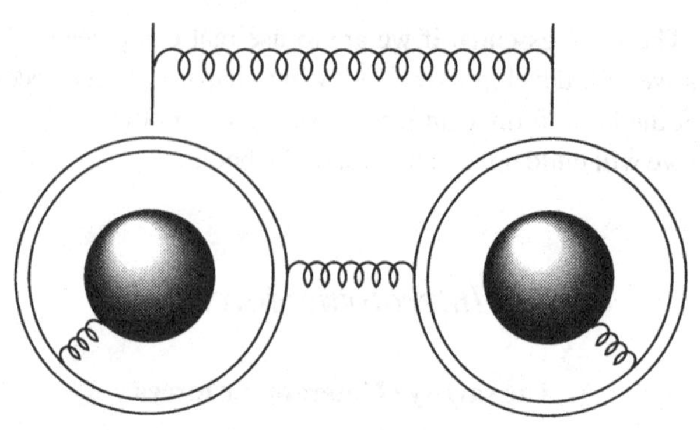

Figure 1.1: Ionic interactions in the shell model.

core(2)–*shell*(2), *core*(1)–*shell*(2), and *core*(2)–*shell*(1). These are illustrated in Figure 1.1.

Van der Waals interactions

The van der Waals interactions are indirect, or second-order forces, in that they occur in addition to the normal electrostatic forces. They are the principal binding forces in solids where the atoms or molecules are electrically neutral (e.g. argon, molecular sulphur, molecular nitrogen). The most important term of this sort gives an interaction energy that is inversely proportional to the sixth power of the interatomic distance, and this is known (for historical reasons) as the *dispersive interaction.* Although on average the atomic charge is spherically distributed around the atom, so that it has no average electrostatic multipole moments, in practice the charge distribution is always fluctuating, giving rise to the existence of temporary moments. The direct interaction between the temporary moments on two atoms will average to zero, but a moment on one atom will induce a moment on its neighbouring atom, and it is the interaction involving this induced moment that does not average to zero. The r^{-6} interaction involves a fluctuating dipole moment on one atom which induces a dipole moment in a second atom. There are also interactions involving higher-order fluctuating and induced moments, which accordingly give energies that have inverse powers higher than 6.

We can gain some insight into the origin of this interaction by considering a simple classical shell model for two neutral atoms. Because the positions of the core and shell for each atom continuously fluctuate, there always exists a non-zero instantaneous value of the electrical dipole moment on each atom. We label the dipole moments of the two atoms p_1 and p_2 respectively. The electrostatic energy between the two atoms, neglecting orientational components,[1] is simply proportional to p_1p_2/r^3. The mean values of these dipole moments over any period of time are zero: $\langle p_1 \rangle = \langle p_2 \rangle = 0$. Therefore if these two moments are uncorrelated, the average value of the electrostatic interaction will also be zero. However, both the dipole moments will generate instantaneous electric fields. If the distance between the atoms is r, the field seen by atom 2, E_2, will be

$$E_2 \propto p_1 / r^3 \tag{1.3}$$

This field will induce an instantaneous extra moment in atom 2:

$$\Delta p_2 \propto E_2 \propto p_1 / r^3 \tag{1.4}$$

This will give an additional energy:

$$\varphi(r) \propto -p_1 \Delta p_2 / r^3 \propto -p_1^2 / r^6 \tag{1.5}$$

which has a non-zero average value, and which is always attractive. In a rough way this demonstrates the origin of the van der Waals r^{-6} attractive interaction.[2] It should be noted that the dipolar fluctuations in the electron distribution are considerably faster than the normal atomic motions. There are also higher-order fluctuating moments, which give rise to higher-order interactions that vary as r^{-n}, where $n > 6$. The general van der Waals energy can thus be expressed as

$$\varphi(r) = -\sum_n A_n r^{-n} \tag{1.6}$$

where $A_n = 0$ for $n < 6$, and it turns out that $A_7 = A_9 = 0$ also. The property of pairwise additivity (see below) holds only for $n = 6$ and $n = 8$. However, in general $A_8 r^{-8} \ll A_6 r^{-6}$, so it is usual to neglect the terms for $n > 6$.

Repulsive forces

When two atoms get sufficiently close their respective electron distributions start to overlap. This overlap has two effects: one is that electrons in the two atoms start to interact with each other directly rather than indirectly via interactions between the complete atoms, and the other is that the Pauli exclusion principle requires that the electrons in the overlapping region should jump into higher energy states. These two effects raise the energy rapidly with increasing

[1] The orientational dependence does not affect the proportionality.

[2] This is a classical argument. The quantum-mechanical theory was first provided by London (1930), and a simplified description is given by Kittel (1976, pp 78–79).

overlap, thereby giving a net repulsion that is short-ranged. Although it is possible to calculate the repulsive interaction from first principles, it is common to use a functional form of the repulsive interaction potential with model parameters. One typically used function, which involves the interatomic distance r, is Br^{-n} (where n is often 12). The r^{-12} repulsion is sometimes combined with the dispersive interaction in a functional form known as the *Lennard-Jones potential*:

$$\varphi(r) = -4\varepsilon\left[\left(\frac{\sigma}{r}\right)^6 - \left(\frac{\sigma}{r}\right)^{12}\right] \tag{1.7}$$

It is easy to show that when two atoms interact via the Lennard-Jones potential, ε is equal to the potential energy at the equilibrium separation, and σ is the distance between the atoms at which the energy is equal to zero.

Another commonly used function for the repulsive interaction is an exponential term known as the *Born–Mayer* interaction (Born and Mayer 1932):

$$\varphi(r) = B\exp(-r/\rho) \tag{1.8}$$

The parameters B and ρ are usually determined empirically, although there is some theoretical justification for the exponential repulsion and the parameters can be calculated using quantum-mechanical methods (e.g. Post and Burnham 1986). ρ is related to the relative sizes of the atoms, and B is a measure of the hardness of the interaction. The combination with the r^{-6} dispersive interaction is known as the *Buckingham potential*, which is commonly used in many different systems with and without electrostatic interactions.

It is worth pointing out that a large amount of work has been carried out in which it has been assumed that atoms have effectively infinitely hard surfaces, like billiard balls or ball bearings. This model – called the *hard sphere model* – can sometimes give results that are surprisingly close to reality!

Metallic and covalent bonding

In metals the atomic cores (nuclei plus the tightly bound inner electrons) are surrounded by a more-or-less uniform density of free electrons. It is this general distribution of electrons that gives metals their electrical conductivity, and the electrons also contribute significantly to the high thermal conductivity. On the other hand, crystals in which the atoms are held together by covalent bonds (such as diamond) prove to be good electrical and thermal insulators. The common feature is that in neither type are the electrons that are important for the cohesion of the crystal localised around the cores of the atoms as they are in an ionic or molecular crystal. This means that it is difficult to calculate the forces between atoms without taking into account the way the electron density changes in response to these forces. Thus the motions of atoms in these systems are accompanied by significant changes in the surrounding electron distribution. It is possible to calculate the forces between atoms in these types of solid, but these calculations are not easy and techniques for such calculations are still topics of current research activity. For example, theoretical calculations of the interactions in silica (SiO_2) are being carried out using the quantum mechanics of small clusters (Lasaga and Gibbs 1987, 1988; Tsuneyuki et al. 1988; Kramer et al. 1991) or of ideal structures (Cohen 1991; Lazarev and Mirgorodsky 1991; McMillan and Hess 1990; Nada et al. 1990). It is possible to use approximate model potentials for covalent systems. For example, it turns out that silicate minerals can be modelled surprisingly accurately using simple model interactions (e.g. Buckingham interactions) for cation–oxygen and oxygen–oxygen interactions, together with the normal Coulombic interactions (Burnham 1990; Leinenweber and Navrotsky 1988; Stixrude and Bukowinski 1988). Refined empirical models also include shell-model interactions for the oxygen ions and O–Si–O bond-bending interactions that vary with the bond angle θ:

$$\varphi(\theta) = \frac{1}{2}K(\theta - \theta_0)^2 \tag{1.9}$$

where θ_0 is the equilibrium bond angle, equal to 109.47° (= $\cos^{-1}(-1/3)$) for tetrahedral angles and 90° for octahedral coordination (Sanders et al. 1984).

Additional comments

In this book, we will be using two approximations throughout. Our first is that the energy of a system of three or more atoms can be represented as the sum of interactions between the different pairs. This is called *pairwise additivity*. An illustration of this is that the energy of the sun–earth–moon system is the sum of the gravitational energies between the sun and earth, the sun and moon, and the earth and moon. Our second approximation is to assume that because the electrons move much faster than the atom cores, the electrons are always in an equilibrium configuration when the atom cores are moving (this approximation is known as the *adiabatic* or *Born–Oppenheimer approximation* (Born and Oppenheimer 1927)). This is particularly relevant for covalent and metallic systems, but we shall not say anything more on this point.

One might be tempted to think that the forces between atoms are reasonably well understood nowadays. Unfortunately this is not the case, as is evidenced by the large number of recent studies of the interactions in quartz (Sanders et al. 1984; Burnham 1990; Leinenweber and Navrotsky 1988; Stixrude and Bukowinski 1988; Lasaga and Gibbs 1987, 1988; Tsuneyaki et al. 1988; Kramer et al. 1991; Lazarev and Mirgorodsky 1991; McMillan and Hess 1990; Nada et al. 1990). It generally turns out to be extremely difficult to calculate the parameters used in the equations given above, and it is nearly always more accurate to derive the parameter values empirically than from calculations. The limitations of the models or fitting procedures mean that even when good agreement is obtained with one set of data (e.g. crystal structure) the model may not reproduce another set of data (e.g. elastic constants). Furthermore, it should be noted that very few substances actually fit neatly into the categories outlined above. Silicates again provide an example. The nearest-neighbour Si–O interactions are partially covalent, but there will always be significant residual charges on the Si and O atoms. Thus a model for silicates must include both ionic and covalent interactions. Given that the oxygen ion is highly polarisable, a good model should also include van der Waals interactions and a shell model. However, it is found that it is possible to model many silicates using only Coulombic interactions and short range Born–Mayer interactions. We will carry on in spite of all such difficulties, but we should always bear in mind the fact that our models for interatomic forces will at best be only crude representations of reality, even when we get very good agreement with experimental results!

Lattice energy

The total energy of a crystal is the sum over all the individual atom–atom interactions (within the assumption of pairwise additivity). Thus if φ_{ij} is the energy between any two atoms i and j, the full energy of the crystal (the *lattice energy*), W, is

$$W = \frac{1}{2}\sum_{i,j} \varphi_{ij}\left(r_{ij}\right) \tag{1.10}$$

where we have assumed that φ_{ij} is a simple function of the separation distance r_{ij}. The factor of $\frac{1}{2}$ in equation (1.10) arises from the fact that the summation in equation (1.10) involves counting each interaction twice!

The summation in equation (1.10) technically includes interactions between all atoms in the crystal. Such a summation is computationally impossible. For interactions that fall off rapidly with distance (e.g. the Lennard-Jones or Buckingham interactions) the summation can be restricted to atoms that are closer than a pre-determined limit (typically 5–10 Å), since the terms for larger distances will be negligibly small. Cut-off limits cannot be used for Coulombic interactions, since the summation does not converge on increasing the interaction distance. For these interactions more complex mathematical techniques are required in order to evaluate the lattice sums correctly. One such technique, the *Ewald Sum*, is described in Appendix A.

The crystal is defined by a set of structural parameters $\{p_l\}$, which includes the unit cell parameters and the coordinates of each atom within the unit cell. When the crystal is in equilibrium, the average force on each atom is zero. At the temperature of absolute zero for a classical crystal (for which there is no motion at $T = 0$ K), this is equivalent to the conditions

$$\frac{\partial W}{\partial p_l} = 0 \quad \text{for all parameters } p_l \tag{1.11}$$

If we take, for example, the unit cell parameter a, the equilibrium condition gives

$$\frac{\partial W}{\partial a} = \frac{1}{2}\sum_{i,j} \frac{\partial \varphi_{ij}}{\partial r_{ij}} \frac{\partial r_{ij}}{\partial a} = 0 \tag{1.12}$$

The solution of the set of equations (1.11) gives the set of structural parameters $\{p_l\}$ (Williams 1972). For complex structures this is a problem that has to be solved on a computer. Although the condition (1.11) is strictly only applicable at $T = 0$ K, most people who work on modelling of crystals use the concept of the equilibrium lattice without worrying about temperature, mainly for reasons of convenience. For many applications, the errors introduced by neglecting temperature in the development of a model are not significant compared with the errors inherent in the model interatomic potential, and are therefore not always worth the effort trying to avoid. For certain applications though, for example for the prediction of thermal expansion, the effects of temperature must be fully considered, and a method based on the free energy rather than the lattice energy is described in Chapter 5.

Many of the models that we will consider in this book contain a number of phenomenological parameters such as the parameters B and ρ in equation (1.8). The concept of the equilibrium lattice can be used with modelling methods to obtain best estimates of these parameters. If P_m represents a parameter in a model interatomic potential, the best estimate of its value is obtained using a minimisation procedure, such that P_m is found as a solution to the equation

$$\frac{\partial}{\partial P_m} \sum_l \left(\frac{\partial W}{\partial p_l} \right)^2 = 0 \tag{1.13}$$

where W is evaluated using the crystal structure determined experimentally. In this equation, the sum is over all the structural parameters p_l. The best set of parameters, $\{P_m\}$, is that for which the sum

$$M = \sum_l \left(\frac{\partial W}{\partial p_l} \right)^2 \tag{1.14}$$

is at a global minimum value.

Worked example: a simple model for NaCl

Consider the crystal structure of NaCl. There is only one structural parameter, namely the unit cell parameter a, with the value of 5.64 Å. The lattice energy has been measured as –764.4 kJ mol^{-1} (data for NaCl have been taken from a compilation given in Kittel 1976, p 92). We can try using a simple model for this crystal. We will assume, quite reasonably, that the ions have their formal charges (+1 electronic charge for Na, –1 electronic charge for Cl). The total Coulomb energy needs to be evaluated using the Ewald sum, but for systems with simple cubic structures such as the NaCl structure the calculated Coulomb energy, W_C, can be expressed in a simple manner due to Madelung (1918):

$$W_C = \frac{N_A e^2}{4\pi\varepsilon_0} \sum_j (\pm)_j r_j^{-1} = -\frac{N_A e^2}{4\pi\varepsilon_0} \frac{2\alpha}{a} \tag{1.15}$$

where the alternative signs account for the interactions between like and unlike charges. The sum is over all atoms in the crystal with respect to a *single* reference atom, and the factor of $\frac{1}{2}$ in equation (1.10) is cancelled by the fact that there are two atoms in the asymmetric unit. The constant α in equation (1.15) is called the *Madelung constant*,[3] and for the NaCl lattice it has the value α = 1.7476.

We need to add to this model a repulsive interaction between nearest-neighbour Na and Cl atoms, for which we will choose an exponential term as given by equation (1.8). Since each atom is surrounded by 6 neighbours at distance $a/2$, and there are 2 atom types, the total contribution to the lattice energy from the repulsive interactions, W_R, is

$$W_R = 6B\exp(-a/2\rho) \tag{1.16}$$

[3] The Madelung constant is a simple way of expressing the Ewald summation (Appendix A) for simple cubic crystals, as described in Kittel (1976, pp 86–91).

where B and ρ are constants that need to be determined empirically, and B has the units of energy per mole. The total lattice energy, W, is therefore given as

$$W = W_{\mathrm{C}} + W_{\mathrm{R}} \tag{1.17}$$

When we substitute the value for a in equation (1.15) we obtain the value $W_{\mathrm{C}} = -861.0$ kJ mol^{-1}. We then obtain $W_{\mathrm{R}} = 96.6$ kJ mol^{-1} from equation (1.17).

The condition for equilibrium is that $\partial W/\partial a = 0$. We therefore have

$$\frac{\partial W}{\partial a} = -\frac{W_{\mathrm{C}}}{a} - \frac{W_{\mathrm{R}}}{2\rho} = 0 \tag{1.18}$$

Substitution of our values for W_{C}, W_{R}, and a enables us to obtain the value $\rho = 0.3164$ Å. We can now substitute these values into equation (1.16) to obtain the value $B = 1.1959 \times 10^5$ kJ mol^{-1}. This completes the development of the model.

We now need to test our model against further experimental data. Let us take the *bulk modulus*, K, defined as

$$K = V\frac{\partial^2 W}{\partial V^2} \tag{1.19}$$

where V is the equilibrium volume for 1 mole of NaCl ion pairs ($= N_{\mathrm{A}}a^3/4$). The bulk modulus is a measure of the resistance of a crystal against compression, and for NaCl it has the value $K = 2.4 \times 10^{10}$ N m^{-2}. We can use the partial differential result,

$$\frac{\partial}{\partial V} = \left(\frac{\partial V}{\partial a}\right)^{-1}\frac{\partial}{\partial a} = \frac{4}{3N_{\mathrm{A}}a^2}\frac{\partial}{\partial a} \tag{1.20}$$

to show that the bulk modulus can be expressed as

$$K = \frac{4}{9N_{\mathrm{A}}a}\frac{\partial^2 W}{\partial a^2} \tag{1.21}$$

Noting that

$$\frac{\partial^2 W}{\partial a^2} = \frac{2W_{\mathrm{C}}}{a^2} + \frac{W_{\mathrm{R}}}{4\rho^2} \tag{1.22}$$

we can calculate a value for $K = 2.45 \times 10^{10}$ N m^{-2}. This compares rather nicely with the experimental value given above, and indicates that our simple model, which was derived only from the measured value of the lattice energy and the equilibrium unit cell length, has some general applicability. More sophisticated models would include shell-model interactions and dispersive interactions, and would relax the assumption of formal charges. Parameterisation of a more complicated model would require the use of additional data, such as the elastic constants, dielectric constants at high and low frequencies, and vibrational frequencies (Sangster et al. 1978; Sangster and Atwood 1978).

This simple example illustrates the methods by which phenomenological models may be developed. For more complex systems with several atom types and more sophisticated models, the number of parameter values that will need to be refined will be so large that computer calculations will be essential. Our example has also illustrated the fact that the simple models described in this chapter can reproduce experimental data surprisingly well.

Transferable models

There is a considerable range of crystal structures based on the elements carbon and silicon, which always bond covalently. Organic crystals are generally composed of molecules as discrete units, which are held together with the van der Waals and higher-order multipolar forces that are much weaker than the forces involved in the intramolecular bonding. For this reason the melting points of many organic crystals are relatively low. On the other hand, silicate crystals are often formed as semi-infinite framework structures, with SiO_4 units as the basic building blocks that are connected together by the corner sharing oxygen atoms. The existence of the framework means that the whole structure is held together by covalent bonds. Other silicates have isolated SiO_4 or Si_2O_7 molecular units, which are electrically charged and therefore bind with the strong Coulomb energy. The common features of both types of system mean that it is possible to develop models that can be applied to any structure: such a model is called a *transferable* model.

Organic crystals

The idea of transferable models was first developed in the 1960s for crystals containing hydrocarbon molecules (Williams 1966, 1967). It was reasonably assumed that the relevant interactions can be represented as Buckingham interactions involving the carbon and hydrogen atoms. A set of potential parameters for C–C, C–H and H–H interactions was obtained using the conditions (1.13) and (1.14) for a reasonably large database of crystal structures. It was quickly appreciated that the same potential set could be used for both aliphatic and aromatic molecules. The basic model has been developed by adding constraints between the parameters; for example, the dispersive interaction $-Ar^{-6}$

is usually subject to the constraint that $A_{CH} = (A_{CC}A_{HH})^{1/2}$, which follows from the fact that the dispersive interaction is proportional to the product of the polarisabilities of the two interacting atoms. The model has also been developed to allow for the existence of small charges on the atoms, and it has been extended to include N, O, F and Cl atoms (Williams 1973; Williams and Cox 1984; Williams and Houpt 1986; Hsu and Williams 1980; Cox et al. 1981). Extensive databases of crystal structures are used for the development of these models. The models have had considerable application in the study of organic crystals and polymers, and are used routinely for drug development.

Silicates

More recently the idea of transferability has been applied to silicates, although the development of appropriate models has not followed as rigorous a path as for the models for organic crystals. In general models have been developed for quartz and then applied to other systems. A range of models has been used. Some models assume formal charges ($4e$ for Si, $-2e$ for O etc., Burnham 1990; Sanders et al. 1984) whilst other models account for the charge redistribution through the covalent bonds by the use of partial charges. The simplest models just use Born–Mayer repulsive interactions in addition to the electrostatic interactions, effectively treating the silicate crystal as an ionic crystal. A useful set of parameters for these interactions has been obtained by *ab initio* calculations[4] (Post and Burnham 1986). The most sophisticated empirical model uses Buckingham interactions for Si–O and O–O interactions, bond-bending O–Si–O interactions of the form of equation (1.9) to account for the covalent nature of the Si bonding, and a shell model for the oxygen atoms to account for the relatively high polarisability of the O^{2-} ion. This model was developed by fitting against the structure and lattice dynamics of quartz (Sanders et al. 1984). The model has been extended for other silicates containing aluminium and additional cations by including additional Born–Mayer cation–oxygen interactions, sometimes with parameters that have been obtained from *ab initio* calculations. The transferable nature of these models has been well-established by application to a wide range of silicates, thereby allowing these models to be used as predictive tools (Price et al. 1987a,b; Jackson and Catlow 1988; Dove 1989; Purton and Catlow 1990; Winkler et al. 1991a; Patel et al. 1991).

[4] *ab initio* calculations are exact calculations that do not use any experimental data. However, the use of approximations in the methods that facilitate such calculations may lead to inaccurate results.

Waves in crystals

The wave equation

The equation of a travelling wave in one dimension is given, in complex exponential form, by

$$u(x,t) = \tilde{u}\exp(i[kx - \omega t]) \tag{1.23}$$

where

u = dynamic variable (such as a displacement) that is modulated in space, x, and time, t
$\tilde{u}$ = amplitude
k = wave vector = $2\pi/\lambda$, where λ = wavelength
ω = angular frequency = $2\pi \times$ frequency

We prefer to work with the complex exponential rather than a single sine or cosine because it gives the easiest representation for further manipulation, although both solutions are perfectly acceptable. It should be noted that we have subsumed all factors of 2π into the constants ω and k, which is the usual practice of physicists if not of crystallographers! Equation (1.23) is a solution of the general wave equation:

$$\frac{\partial^2 u}{\partial t^2} = c^2 \frac{\partial^2 u}{\partial x^2} \; ; \; c = \frac{\omega}{k} \tag{1.24}$$

The solution $u(x, t)$ is a sinusoidal function with constant wavelength. The motions described by this function are called *harmonic* motions. Any point along x will vibrate with angular frequency ω. The form of equation (1.24) is such that the wave will maintain its sinusoidal form with constant amplitude for all times, but the positions of the maxima and minima will change with time. $u(x, t)$ is therefore a travelling wave rather than a standing wave. The position of one of the maxima (for which $x_{max} = 0$ at $t = 0$) moves so that the exponent in equation (1.23) remains at zero. The position of this maximum, x_{max}, changes as

$$x_{max} = \frac{\omega t}{k} \tag{1.25}$$

which corresponds to the maximum (or peak) moving with a constant velocity c given as $c = \omega/k$ (the same parameter c as in equation (1.24)). This velocity is called the *phase velocity*; all the peaks and troughs in the wave move at this constant velocity. We can also define another velocity, called the *group velocity*, which is given by $\partial\omega/\partial k$. The group velocity gives the velocity of a wave packet composed of a narrow distribution of frequencies about a mean value ω.

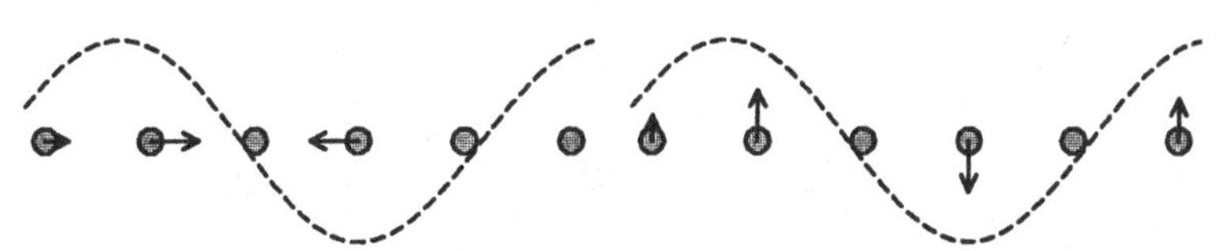

Figure 1.2: Longitudinal (left) and transverse (right) waves in a one-dimenstional crystal.

The group velocity also gives the velocity of the energy flow associated with a travelling wave. A standing wave will have a zero value of the group velocity.

Travelling waves in crystals

How can we visualise this wave travelling though a crystal, where the space that vibrates is not continuous (like a string on a musical instrument) but is composed of discrete atoms? The answer is to think of our wave as representing displacements, $u(x, t)$, of the atoms from their equilibrium position. These displacements are shown in Figure 1.2 for *longitudinal* (i.e. compressional) and *transverse* (i.e. perpendicular) vibrations.

In three dimensions the travelling wave in the crystal gets a bit difficult to visualise, so we will consider the two-dimensional case in more detail first. A simple wave is shown in Figure 1.3, where we mark the positions of the maxima and minima (continuous and broken lines respectively), and the longitudinal displacements of the atoms. The wavelength λ is also indicated. The wave travels in the direction normal to the lines of maxima. We can now extend our definition of the wave vector k so that it contains information about both the wavelength and the propagation direction of the waves. The wave vector then becomes the vector $\mathbf{k}$ pointing in the direction of propagation with modulus $2\pi/\lambda$. Since the lines of maxima really outline planes of atoms, and the normals to planes give vectors in reciprocal space, the wave vector $\mathbf{k}$ defines a vector in reciprocal space. This is consistent with its definition, with units of inverse length.

The generalisation to three dimensions is trivial, but not so easy to visualise. The wave vector $\mathbf{k}$ now has three components. The atomic displacements associated with a wave, $\mathbf{u}(\mathbf{r}, t)$, where $\mathbf{r}$ is the equilibrium position of an atom, are similarly three-dimensional vectors. These displacement vectors may be parallel to $\mathbf{k}$ (*longitudinal*), perpendicular to $\mathbf{k}$ (*transverse*), or, in the general case, along a direction that is not directly related to the direction of $\mathbf{k}$. The equation of the wave is thus:

$$\mathbf{u}(\mathbf{r},\ t) = \tilde{\mathbf{u}} \exp\left(i[\mathbf{k}\cdot\mathbf{r} - \omega t]\right) \tag{1.26}$$

where $\tilde{\mathbf{u}}$ is the amplitude of the wave, and is itself a function of the wave vector $\mathbf{k}$. The angular frequency ω is also a function of $\mathbf{k}$.

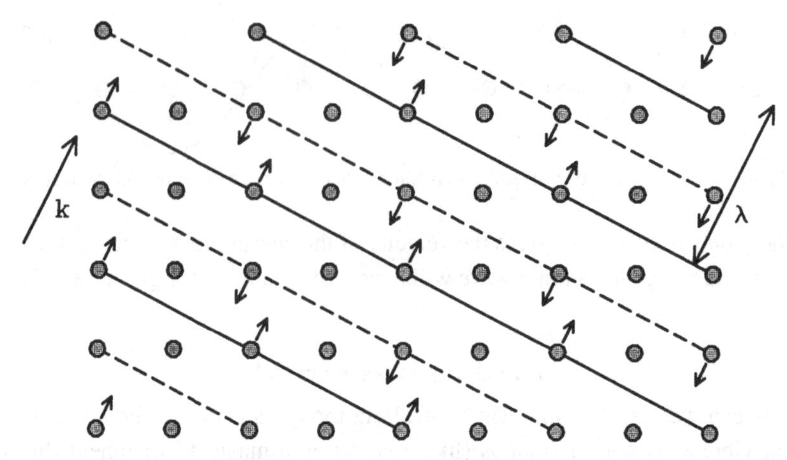

Figure 1.3: Example of a travelling wave in a two-dimensional monatomic crystal. Arrows indicate direction of motion, and planes of atoms are indicated by connecting lines. The wavelength (λ) is equal to two plane separations as indicated. **k** denotes the direction of the wave vector, normal to the planes.

What happens when we have more than one atom in the unit cell? In the one-dimensional transverse case, we can draw out two vibrational modes for a given wave vector; this is shown in Figure 1.4. For one mode, the A and B atoms move together in phase, and for the other mode they move out of phase. How far they move and the frequencies of the two modes are determined by the forces at play (the rest of the book deals with this!). For more complex cases more complicated behaviour can emerge. We will find later on that the following rules apply for a three-dimensional crystal with n atoms per unit cell:

(*a*) There are $3n$ different combinations of motion.
(*b*) Each combination will have a unique frequency at a given wave vector.
(*c*) The direction in which any atom moves due to the wave motion will be determined by the interatomic forces. The amplitude of the motion will be determined by the frequency of the wave and the temperature.
(*d*) For any single wave, each atom moves with the same frequency and wave vector.
(*e*) The crystal can (and in general will) contain all possible modes of vibration as a linear superposition at any time.

This discussion sets the scene. We now need to consider three questions:

(*a*) What are the allowed values of the wave vectors?
(*b*) What are the values of the frequencies?
(*c*) What are the values for the amplitudes of the waves?

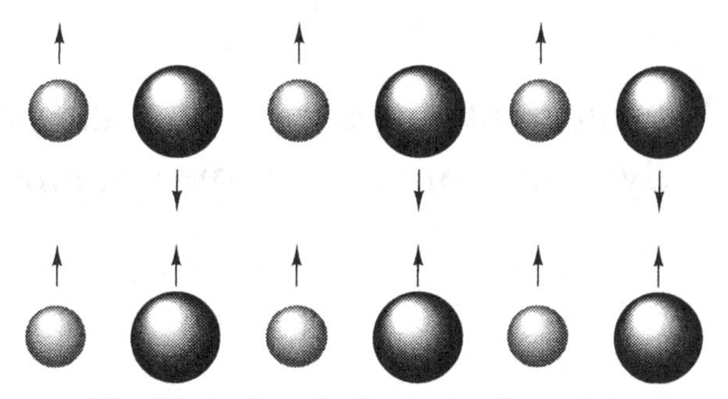

Figure 1.4: Two possible vibrational modes for a unit cell containing two atoms. The example shows vibrations with zero wave vector (infinite wavelength).

These questions will be tackled in the next few chapters.

Summary

1 Failures of the static lattice model have been enumerated, highlighting the need for a dynamic model.

2 The different types of interatomic interactions (electrostatic, van der Waals, repulsive, covalent and metallic) have been reviewed.

3 The lattice energy has been defined as the sum over all the individual interatomic interactions within the crystal. The lattice energy can be used in the development of a model potential.

4 The wave equation has been introduced, and its effect on a crystal lattice has been described.

FURTHER READING

Ashcroft and Mermin (1976) ch. 19–21
Kittel (1976) ch. 3

2

The harmonic approximation and lattice dynamics of very simple systems

A simple model for a monatomic crystal is described, and the first results of the theory of lattice dynamics are obtained. The simple model forces us to encounter some of the important general concepts that will be used throughout this book. The chapter concludes with a detailed study of the lattice dynamics of the rare gas crystals.

The harmonic approximation

The first thing we need is a simple model to use as our starting point. The simplest model of all is a linear chain of atoms, each of mass m, and separated by the unit cell length a, as illustrated in Figure 2.1.[1] For the moment we consider that each atom only feels the force of its immediate neighbour, calling this the *nearest-neighbour interaction*. If the energy between two neighbours at a distance of a is $\varphi(a)$, the total energy of a chain of N atoms[2] when each atom is at rest is:

$$E = N\varphi(a) \tag{2.1}$$

Now we assume that each atom can move about a little, and we represent the displacement of an atom along the chain by the symbol u. If the displacements are small in comparison with a, then we can calculate the energy of this flexible chain using a Taylor series, summing over all the atoms:

$$E = N\varphi + \sum_{s\geq 1} \frac{1}{s!}\frac{\partial^s \varphi}{\partial u^s} \sum_n \left(u_n - u_{n+1}\right)^s \tag{2.2}$$

[1] It is just as well to know that strictly speaking the one-dimensional chain is unstable, as is also a two-dimensional plane! Essentially the chain will shake itself to pieces unless it is stabilised by contact with a three-dimensional object such as an adsorbed monolayer on a crystal surface.
[2] Technically we have joined the ends of the chain in order to remove the problem that the atoms at the ends of the chain only have one neighbour each.

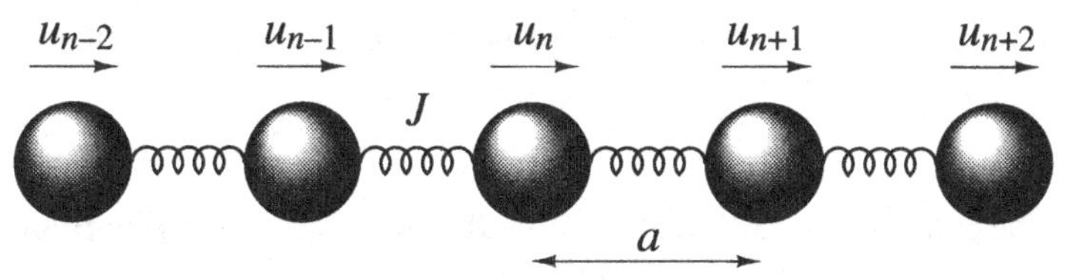

Figure 2.1: Linear chain model. J is the harmonic force constant (the interactions are represented by springs), and the atomic displacements are represented by u.

If u_n is the displacement of the n-th atom from its equilibrium position, the distance between two atoms n and $n + 1$ is $r = a + (u_n - u_{n+1})$. Thus the derivatives of φ with respect to u are equivalent to the derivatives with respect to r. Since a is the *equilibrium* unit cell length, the first derivative of φ is zero, so that the linear term in the expansion ($s = 1$) can be dropped. All the other differentials correspond to the point $u = 0$. As u is small in comparison to a, this series is a convergent one, and we might expect that the dominant contribution will be the term that is quadratic in u. Therefore we start with a model that includes only this term ($s = 2$), neglecting all the higher-order terms. The energy of this lattice is then the same as the energy of a set of harmonic oscillators, and so we call this approximation the *harmonic approximation*. The higher-order terms that we have neglected are called the *anharmonic terms*.

Why do we make this approximation? The main reason is that it is a mathematically convenient approximation. We know that the harmonic equations of motion have exact solutions, whereas even the simplest anharmonic equations do not have exact solutions but require the use of approximation schemes. Although this may not sound a very noble reason, the use of this approximation can be justified in at least three related ways. Firstly, as will be seen, the harmonic approximation in practice proves to be capable of giving good results. The anharmonic part of the model usually leads to only a small modification of the overall behaviour, and since the amplitudes of the displacements will be expected to decrease on lowering the temperature (i.e. the kinetic energy of the chain) the harmonic term will be the only important term at low temperatures. Secondly, the harmonic approximation allows us to obtain many of the important physical principles characteristic of the system with only a minimum of effort, and it is these that we are hoping to study. Thirdly, there is an often-used approach in physics when dealing with complex problems, which is to solve the simple model first (in this case the harmonic model), and then correct the simple solution for the more complex parts (the anharmonic corrections). This is called the *perturbation method*. In Chapter 8 we shall see that there are important features of real crystals that the harmonic approximation fails to explain, but we will be able to progress by simply correcting the harmonic results without having to start again!

The equation of motion of the one-dimensional monatomic chain

The harmonic energy of our chain from equation (2.2) is[3]

$$E^{\mathrm{harm}} = \frac{1}{2} J \sum_n \left(u_n - u_{n+1}\right)^2 \ ; \ J = \frac{\partial^2 \varphi}{\partial u^2} \tag{2.3}$$

The equation of motion of the n-th atom from the classical Newton equation is then

$$m\frac{\partial^2 u_n}{\partial t^2} = -\frac{\partial E^{\mathrm{harm}}}{\partial u_n} = -J\left(2u_n - u_{n+1} - u_{n-1}\right) \tag{2.4}$$

If our chain contains N atoms, we need to think about what happens to the ends of the chain. The usual trick for a long chain is to join the ends; this is called the *Born–von Kármán periodic boundary condition* (Born and von Kármán 1912, 1913; Born and Huang 1954, pp 45–46, App. IV). We know that the solution of the harmonic equation of motion is a sinusoidal wave, so the motion of the whole system will correspond to a set of travelling waves as given by equation (1.23). Our aim then is to find the set of frequencies of these waves. We expect the time-dependent motion of the n-th atom to be a linear superposition of each of the travelling waves allowed along the chain; the mathematical representation is:

$$u_n(t) = \sum_k \tilde{u}_k \exp\left(i\left[kx - \omega_k t\right]\right) \tag{2.5}$$

where k is any wave vector (= 2π/wavelength), ω_k is the corresponding angular frequency, $\tilde{u}_k$ is the amplitude, and x is restricted to the values $x = na$. We will find later that there is also a discrete set of allowed values of k. However, if for the moment we consider each wave vector separately, we can solve the equation of motion for each individual wave. We simply substitute equation (2.5) into equation (2.4), to obtain:

[3] Note that this can also be written as $\frac{1}{2}K\sum_n\left(u_n^2 - u_n u_{n+1}\right)$, where $K = 2J = \left(\frac{\partial^2 E}{\partial u_n^2}\right) = \left(\frac{\partial^2 E}{\partial u_n \partial u_{n+1}}\right)$ for all values of n.

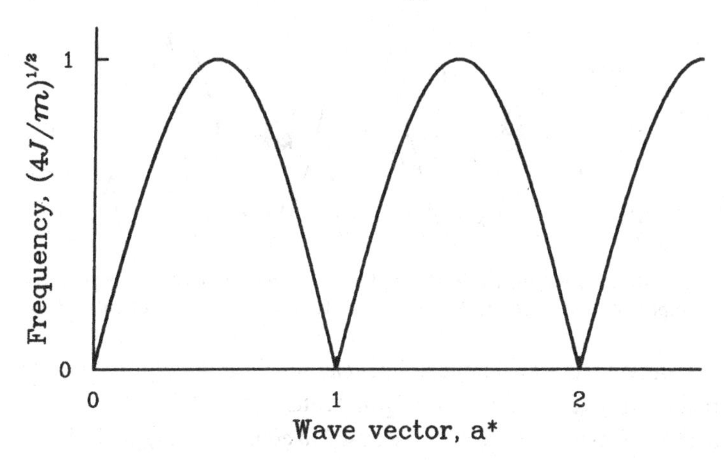

Figure 2.2: Acoustic mode dispersion curve (angular frequency ω *vs.* wave vector k) for the one-dimensional monatomic chain.

$$
\begin{aligned}
&-m\omega_k^2\tilde{u}_k \exp\left(i\left[kna - \omega_k t\right]\right) \\
&= -J\tilde{u}_k \exp\left(-i\omega_k t\right)\left[2\exp(ikna) - \exp\left(ik[n-1]a\right) - \exp\left(ik[n+1]a\right)\right] \\
&= -2J\tilde{u}_k \exp\left(i\left[kna - \omega_k t\right]\right)\left[1 - \frac{\exp(ika) + \exp(-ika)}{2}\right]
\end{aligned}
\qquad (2.6)
$$

If we cancel the constant expressions from both sides of the equation, we obtain an expression for the angular frequency as a function of wave vector, ω_k:

$$
\begin{aligned}
m\omega_k^2 &= 2J(1 - \cos ka) \\
\Rightarrow \omega_k^2 &= \frac{4J}{m}\sin^2(ka/2) \\
\Rightarrow \omega_k &= \left(\frac{4J}{m}\right)^{1/2} \left|\sin(ka/2)\right|
\end{aligned}
\qquad (2.7)
$$

By taking only the positive roots we obtain the behaviour of the angular frequency as shown in Figure 2.2. The graph of ω_k is known as a *dispersion curve.*

Reciprocal lattice, the Brillouin zone, and allowed wave vectors

One striking feature of the dispersion curve shown in Figure 2.2 is the periodicity of the function. For unit cell length a, the repeat period is $2\pi/a$, which is equal to the unit cell length a^* in the reciprocal lattice. Thus our dynamic analysis has directly given us a new lattice in reciprocal space which is equiva-

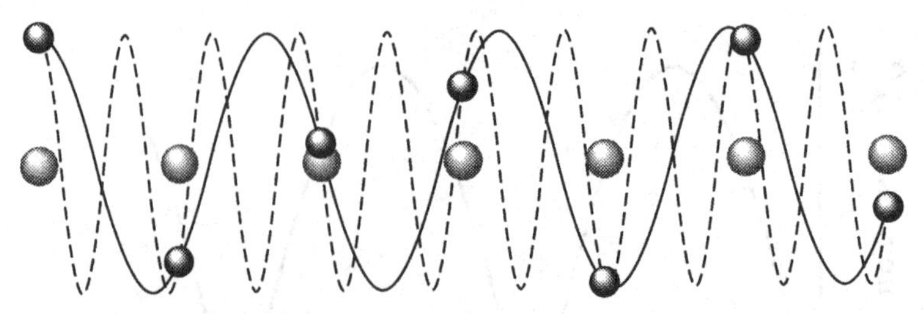

Figure 2.3: The atomic motions associated with two waves that have wave vectors that differ by a reciprocal lattice vector. We see that the two waves have identical effects.

lent to the standard reciprocal lattice. This feature will follow for three-dimensional lattices also, even with non-orthogonal cells.

If we take any wave vector from the set used in equation (2.4), and add to it a reciprocal lattice vector, the equation of motion for this second wave will be identical to that for the original wave. Therefore the only useful information is contained in the waves with wave vectors lying between the limits

$$-\frac{\pi}{a} < k \leq \frac{\pi}{a} \tag{2.8}$$

We call this range of wave vectors the *first Brillouin zone* (Brillouin 1930). The repeated zones are the second, third Brillouin zones etc. These zones will be important when we consider anharmonic interactions in Chapter 8 and neutron scattering in Chapter 9. We note that the wave vectors $k = \pm \pi/a$ define special points in reciprocal space, called *Brillouin zone boundaries*, which lie half-way between reciprocal lattice points. In our example the group velocity, $\partial\omega/\partial k$, is equal to zero at these points, meaning that the wave with this wave vector is a standing wave. It will turn out that this is a general result in three dimensions.

We can easily show that two wave vectors k and k' that differ by a reciprocal lattice vector G, where $k' = k + G$, have an identical effect on the atoms in the chain. We know that $\omega_k = \omega_{k'}$. When $x = na$ we have

$$\exp(ik'x) = \exp(iGx) \times \exp(ikx) = \exp(ikx) \tag{2.9}$$

Thus the travelling wave solution (1.23) is the same for k' as for k. This is illustrated in Figure 2.3, where we see that two waves with wave vectors differing by a reciprocal lattice vector give identical displacements at each lattice position. This result can be generalised to three dimensions, and to unit cells containing more than one atom.

In three dimensions, as in one dimension, the Brillouin zone boundaries are defined as lying half-way between reciprocal lattice points, although now the

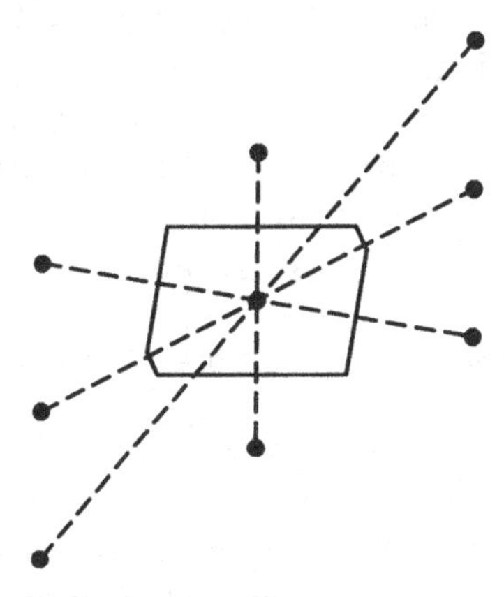

Figure 2.4: Example of the construction of a Brillouin zone in two dimensions.

boundary is a plane rather than a point. The plane boundary is normal to the direction between the two reciprocal lattice points. This is best illustrated in two dimensions, see Figure 2.4. The complete Brillouin zone is the minimum space that is enclosed by a set of boundary planes, and encloses a single reciprocal lattice point at the centre. The Brillouin zone forms a unit that can be packed with other zones to fill the reciprocal space of a crystal completely. In this sense the Brillouin zone is similar to the reciprocal unit cell, although the shape of a Brillouin zone can be more complicated than that of a reciprocal unit cell. The advantage of the Brillouin zone is that the surfaces, and special points on the surfaces, have special symmetry properties, as we have indicated above for our simple chain.

It is common practice to define the wave vector as normalised by the first reciprocal lattice vector lying along the direction of the wave vector. This gives what is called the *reduced wave vector*. For our one-dimensional example, the reduced wave vector has a value of $\frac{1}{2}$ at the Brillouin zone boundary, obtained by dividing the wave vector $a^*/2$ by the reciprocal lattice vector.[4] Thus, in common with most other workers, we will usually show dispersion curves with reduced wave vectors between 0 and $\frac{1}{2}$, noting that for non-primitive unit cells some of the zone boundaries occur with reduced wave vector values of 1. It is also common practice to label special points and lines of symmetry in the Brillouin zone by letters, Roman for points on the surface of the Brillouin zone and Greek for points within the zone. For example, the wave vector $(0, 0, 0)$ is always given the label Γ, and for primitive cubic Brillouin zones the common

[4] This is of course no different from the use of Miller indices for reciprocal lattice vectors.

labels are X for $(\frac{1}{2}, 0, 0)$, M for $(\frac{1}{2}, \frac{1}{2}, 0)$, R for $(\frac{1}{2}, \frac{1}{2}, \frac{1}{2})$, Δ for $[\xi, 0, 0]$, Σ for $[\xi, \xi, 0]$, and Λ for $[\xi, \xi, \xi]$.

Although the dispersion curve is a continuous curve, a finite chain will permit only a discrete set of wave vectors. Since with periodic boundary conditions atom N is identical to atom 0, we have

$$\exp ikNa = \exp 0 = 1 \tag{2.10}$$

Therefore the allowed values of k are given by the discrete set:

$$k = \frac{2\pi m}{Na} \tag{2.11}$$

where m is an integer. This is analogous (but not identical) to the fact that a plucked string will allow only a discrete set of vibrations with integral fractions of the string length as the set of wavelengths. Comparison of equations (2.8) and (2.11) shows that there are N allowed wave vectors within one Brillouin zone. In the general case, the number of allowed wave vectors within one Brillouin zone is equal to the number of primitive unit cells in the crystal.

The long wavelength limit

As k approaches zero, we can take the linear approximation to the dispersion curve given by equation (2.7):

$$\omega(k \to 0) = a\left(\frac{J}{m}\right)^{1/2} |k| \tag{2.12}$$

This gives the phase velocity c, which is equivalent to the velocity of sound in the crystal:

$$c = \frac{\omega}{k} = a\left(\frac{J}{m}\right)^{1/2} \tag{2.13}$$

Because of this relationship to sound waves, a vibrational mode with a dispersion curve of the form given in Figure 2.2, which goes to zero in the limit of small k, is known as an *acoustic mode*. In addition, as the atomic displacements are along the direction of the wave vector, the mode is known as a *longitudinal acoustic mode*.

We can now make a connection with macroscopic elastic properties. If we compress our chain by the strain e, the average distance between atoms is not a but $a' = a(1 - e)$, where $e \ll 1$. The energy of the strained chain (neglecting the energy due to the dynamic atomic motions) therefore becomes equal to:

$$E = N\varphi + \frac{1}{2}NJ(a - a')^2 \tag{2.14}$$

This result follows from the Taylor expansion introduced in equation (2.2), together with our definition of the force constant J given by equation (2.3). Thus the extra energy per atom, called the *strain energy*, is

$$E_{\text{strain}} = \frac{1}{2}J(a - a')^2 = \frac{1}{2}Ja^2e^2 = \frac{1}{2}Ce^2 \tag{2.15}$$

where the constant $C = Ja^2$ is called the *elastic constant*. From the velocity of sound given by equation (2.13) we have

$$c^2 = \omega^2 / k^2 = Ja^2 / m = C / m \tag{2.16}$$

Hence, in the long wavelength limit, we have

$$\omega^2 = Ck^2 / m \tag{2.17}$$

In a three-dimensional crystal the strain energy is defined relative to a unit volume, so that conventionally our expressions (2.16) and (2.17) become

$$C = \rho c^2 \quad ; \quad \rho\omega^2 = Ck^2 \tag{2.18}$$

where ρ is the density. The relationship between the acoustic modes and macroscopic elasticity is examined in more detail in Chapter 7.

Extension to include distant neighbours

Our analysis is readily extended to include more interactions than only the nearest-neighbour interaction. If $\varphi_p(r_p)$ is the energy between the p-th neighbours separated by a distance r_p ($= pa$ for equilibrium), the energy of our chain is given by

$$E = N\sum_p \varphi_p(pa) + \frac{1}{2}\sum_{n,p}\left(\frac{\partial^2\varphi_p}{\partial u^2}\right)_{r_p = pa}\left(u_n - u_{n+p}\right)^2 \tag{2.19}$$

Note that we have again neglected the anharmonic terms. The harmonic energy is therefore given as

$$E^{\text{harm}} = \frac{1}{2}\sum_{n,p} J_p\left(u_n - u_{n+p}\right)^2 \tag{2.20}$$

where

$$J_p = \frac{\partial^2 \varphi_p}{\partial u^2} = \left(\frac{\partial^2 \varphi_p}{\partial r_p^2} \right)_{r_p = pa} \tag{2.21}$$

The equation of motion given by equation (2.4) is readily extended to

$$m \frac{\partial^2 u_n}{\partial t^2} = -\frac{\partial E^{\text{harm}}}{\partial u_n} = -\sum_p J_p \left(2u_n - u_{n-p} - u_{n+p} \right) \tag{2.22}$$

Our travelling wave solution is still given by equation (2.5). Hence the solution of equation (2.22) for each wave vector gives us

$$\begin{aligned} &-m\omega_k^2 \tilde{u}_k \exp\left(i\left[kna - \omega_k t\right]\right) \\ &= -\tilde{u}_k \exp(-i\omega_k t) \sum_p J_p \left[2\exp(ikna) - \exp(ik[n-p]a) - \exp(ik[n+p]a) \right] \end{aligned} \tag{2.23}$$

This simply reduces to

$$\begin{aligned} m\omega_k^2 &= \sum_p J_p \left[2 - \exp(ikpa) - \exp(-ikpa) \right] \\ &= 2\sum_p J_p [1 - \cos kpa] \end{aligned} \tag{2.24}$$

and we finally obtain the expression for the dispersion curve:

$$\omega_k^2 = \frac{4}{m} \sum_p J_p \sin^2\left(\frac{kpa}{2} \right) \tag{2.25}$$

This result represents only a minor modification to the result for nearest neighbours, equation (2.7). The general result has the same behaviour for $k \to 0$ and at $k = \pi/a$ as described above.

Three-dimensional monatomic crystals

If we preserve simplicity, our model can be easily extended to three dimensions, if now in our equations of motion the variables u refer not to displacements of atoms but of planes of atoms, and our force constants are similarly redefined. This was discussed in Chapter 1, and is illustrated in Figure 1.3. We also need to include motions that are perpendicular to the wave vector. These follow as a simple extension of our one-dimensional model. One can imagine

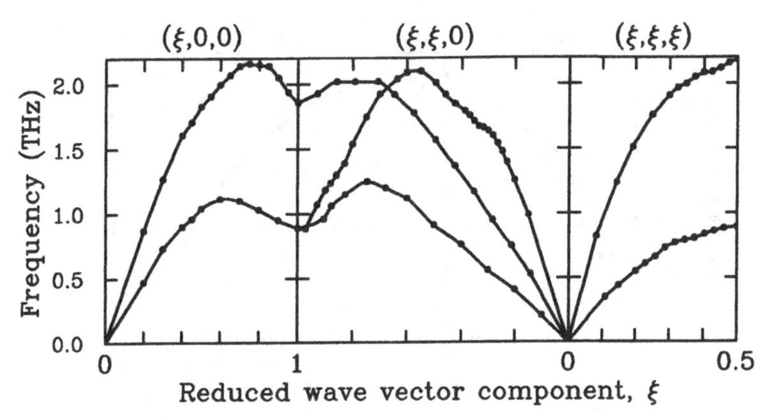

Figure 2.5: Acoustic mode dispersion curves for lead measured by neutron scattering (Brockhouse et al. 1962).

that perpendicular displacements will have different force constants from the longitudinal force constants, but they will be expressed in an identical manner so that the results for the different types of motion will follow identical equations. As there are two orthogonal directions perpendicular to the wave vector, there will be two more acoustic branches with different dispersion curves determined by the new force constants. The corresponding vibrations are called *transverse acoustic* modes. In high-symmetry situations, where there are three-, four- or six-fold rotation axes along the direction of the wave vector, the force constants will be the same for any two transverse displacements, so that the two transverse modes will have degenerate[5] frequency dispersion.

It should be noted, however, that our hand-waving extensions to the simple model are technically appropriate only for crystals of high symmetry and for wave vectors along symmetry directions (e.g. [001]). For more complex cases the force constants between planes are not easily defined, and the motions are not necessarily transverse or longitudinal. We will consider how to do things properly in Chapter 6.

Examples of three-dimensional monatomic unit cells are provided by many metals. Figure 2.5 shows the acoustic mode dispersion curves for lead (fcc structure) for wave vectors along different directions in reciprocal space, and Figure 2.6 shows the acoustic mode dispersion curves for potassium (bcc structure). Note the degeneracies along [100], which is a four-fold rotation axis, and along [111], which is a three-fold rotation axis. The transverse modes are not degenerate along [110], which is only a two-fold rotation axis. The results shown in Figures 2.5 and 2.6 were obtained by neutron scattering

[5] Meaning of exactly equal frequency or energy.

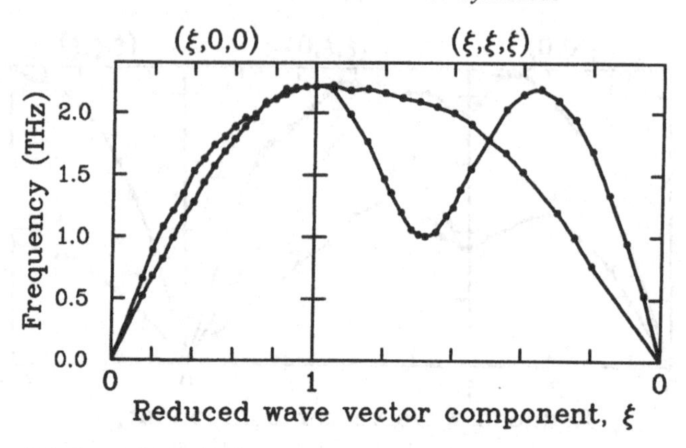

Figure 2.6: Acoustic mode dispersion curves for potassium measured by neutron scattering (Cowley et al. 1966).

experiments. The technique of neutron scattering is described in detail in Chapter 9. For the moment, we can say that the basic idea is that the neutrons are scattered by the lattice vibrations, and the frequency and wave vector of any vibration can be obtained by measuring the change in energy and wave vector of the scattered neutron beam. The change in wave vector corresponds to the wave vector of the lattice vibration, and the change in energy is equal to the frequency of the vibration multiplied by Planck's constant.

Worked example: the lattice dynamics of the rare-gas solids

The rare-gases crystallise with face-centred cubic structures. Table 2.1 lists the atomic masses, melting temperatures T_m, and the unit cell parameters a for the different elements. The measured dispersion curves are given in Figures 2.7–2.10.

The rare-gas elements are good examples for the calculation of lattice dynamics properties since they have simple structures, and the weak forces can be represented by well-defined short-range interactions. Specifically we will assume that the interatomic energy $\varphi(r)$ in each case is given by the standard Lennard-Jones form (equation (1.7)),

$$\varphi(r) = -4\varepsilon\left[\left(\frac{\sigma}{r}\right)^6 - \left(\frac{\sigma}{r}\right)^{12}\right] \tag{2.26}$$

where r is the distance between two atoms. We will assume that only nearest-neighbour interactions need to be considered. Our aim will be to obtain numerical values for the coefficients ε and σ in the Lennard-Jones model from the

Table 2.1: *Crystal data for the rare-gas solids (Kittel 1976, p 77)*

Atom	Atomic mass	T_m (K)	a (Å)
Ne	20.2	23	4.466
Ar	40.0	84	5.313
Kr	83.8	117	5.656
Xe	113.3	161	6.129

experimental data of Figures 2.7–2.10. This illustrates one of the advantages of the study of lattice dynamics, namely that it is possible to extract direct information concerning the interatomic potentials.

We first consider two atoms at their equilibrium separation distance r_0 in the face-centred cubic crystal, $r_0 = a/\sqrt{2}$. The condition that $\partial\varphi/\partial r = 0$ when $r = r_0$ immediately gives us

$$\sigma = 2^{-2/3} a \tag{2.27}$$

We now obtain the harmonic force constant, K, between a pair of atoms at their equilibrium separation:

$$K = \left(\frac{\partial^2 \varphi}{\partial r^2}\right)_{r=r_0} = -\frac{4\varepsilon}{r_0^2}\left[42\left(\frac{\sigma}{r_0}\right)^6 - 156\left(\frac{\sigma}{r_0}\right)^{12}\right] = \frac{72\varepsilon}{2^{1/3}\sigma^2} \tag{2.28}$$

We will proceed by writing down the dynamical equations in terms of K, and we will then use experimental data to determine values for K.

Consider now the lattice vibrations with wave vectors along [111]. This direction is normal to the (111) planes. Each atom in a (111) plane has three neighbouring atoms in the nearest (111) plane. It is convenient to define a coordinate system for the (111) plane such that z is along [111], x is along $[1\bar{1}0]$, and y is along $[11\bar{2}]$. With this coordinate system, the coordinates of the nearest-neighbour atoms in the next (111) plane are:

$$\frac{a}{\sqrt{6}}\left(\pm\frac{\sqrt{3}}{2}, -\frac{1}{2}, \sqrt{2}\right) \quad ; \quad \frac{a}{\sqrt{6}}\left(0, 1, \sqrt{2}\right) \tag{2.29}$$

For longitudinal and transverse motions we need the following inter-planar force constants:

$$J_L = \sum_j \frac{\partial^2 \varphi}{\partial z_j^2} \quad ; \quad J_T = \sum_j \frac{\partial^2 \varphi}{\partial x_j^2} = \sum_j \frac{\partial^2 \varphi}{\partial y_j^2} \tag{2.30}$$

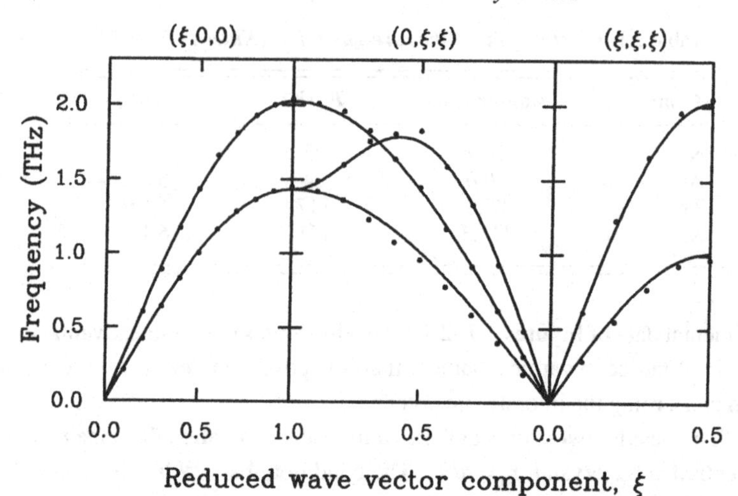

Figure 2.7: Acoustic mode dispersion curves for argon (Fujii et al. 1974). The lines give the calculated dispersion curves using the model described in the text.

where the sum is over the three neighbouring atoms in the next plane. It can easily be shown that

$$\frac{\partial^2 \varphi}{\partial z^2} = \left(\frac{z}{r}\right)^2 \frac{\partial^2 \varphi}{\partial r^2} \quad \text{etc.} \tag{2.31}$$

This gives the result:

$$J_{\mathrm{L}} = K\sum_j \left(\frac{z_j}{r}\right)^2 \quad ; \quad J_{\mathrm{T}} = K\sum_j \left(\frac{x_j}{r}\right)^2 = K\sum_j \left(\frac{y_j}{r}\right)^2 \tag{2.32}$$

where we use the coordinates of equation (2.29), and for all neighbours the nearest-neighbour distance is $r = a/\sqrt{2}$.

Substitution of the coordinates of equation (2.29) into the expressions for the inter-planar force constants, equation (2.32), yields:

$$J_{\mathrm{L}} = 2K \quad ; \quad J_{\mathrm{T}} = K/2 \tag{2.33}$$

From equation (2.7) the zone boundary angular frequencies are given as:

$$\omega_{\mathrm{L}} = \left(\frac{8K}{m}\right)^{1/2} \quad ; \quad \omega_{\mathrm{T}} = \left(\frac{2K}{m}\right)^{1/2} \tag{2.34}$$

These results imply that $\omega_{\mathrm{L}} = 2\omega_{\mathrm{T}}$ for $\mathbf{k} = \frac{1}{2}[111]$, which can be seen from Figures 2.7–2.10 to be approximately true for each example. From the experi-

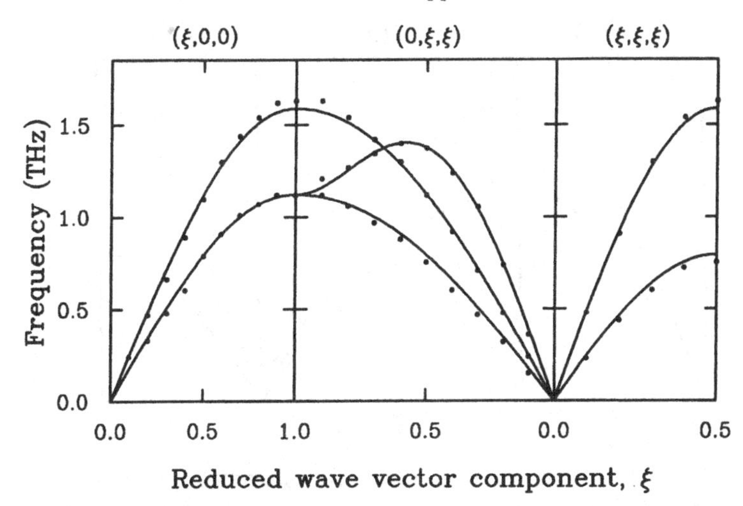

Figure 2.8: Acoustic mode dispersion curves for neon (Endoh et al. 1975). The lines give the calculated dispersion curves using the model described in the text.

mental data we can obtain two estimates for the value of K for each system, which are given in Table 2.2.

Consider now the dispersion curves along [001]. The nearest-neighbour atoms in the next (001) plane from the atom at the origin have coordinates

$$\frac{a}{2}(\pm 1,\ 0,\ 1) \quad ; \quad \frac{a}{2}(0,\ \pm 1,\ 1) \tag{2.35}$$

We do not now need to change coordinate systems as the vibration is along a unit cell vector. Using the results of equations (2.30)–(2.32) we obtain values for the force constants for wave vectors along [001]:

$$J_{\mathrm{L}} = 2K \quad ; \quad J_{\mathrm{T}} = K \tag{2.36}$$

which give zone boundary angular frequencies:

$$\omega_{\mathrm{L}} = \left(\frac{8K}{m}\right)^{1/2} \quad ; \quad \omega_{\mathrm{T}} = \left(\frac{4K}{m}\right)^{1/2} \tag{2.37}$$

The prediction that $\omega_{\mathrm{L}} = \sqrt{2}\omega_{\mathrm{T}}$ for $\mathbf{k} = \frac{1}{2}[001]$ is consistent with the data of Figures 2.7–2.10. We now have another two estimates for the value of K, and all four estimates for each system are compared and averaged in Table 2.2. The largest standard deviation on the average value of K is only 10%, but improved accuracy could have been obtained if we had fitted the curves over the whole branches rather than taking only the zone boundary wave vectors. The errors in

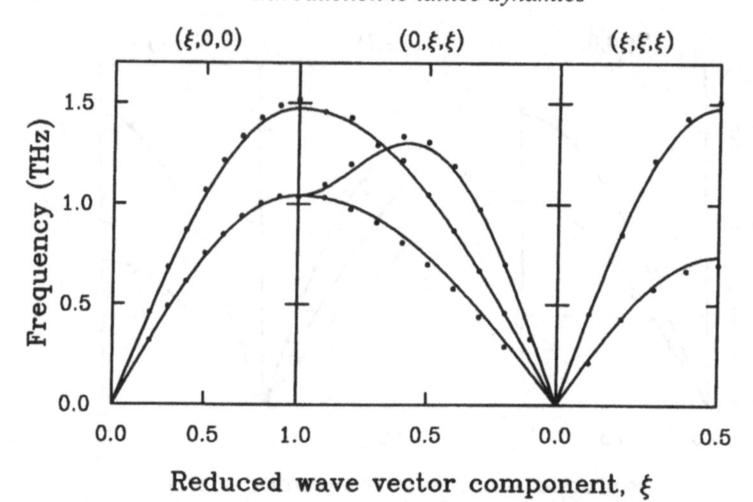

Figure 2.9: Acoustic mode dispersion curves for krypton (Skalyo et al. 1974). The lines give the calculated dispersion curves using the model described in the text.

K are larger than the errors in ω because K is proportional to the square of ω. The average values of K have been used to calculate the complete set of dispersion curves shown as solid lines in Figures 2.7–2.10. The agreement between the calculations and the experimental data is generally reasonable, indicating that our approximation of using only nearest-neighbour force constants is not too far from reality.

The dispersion curves along [011] are slightly more complicated, owing to the fact that nearest-neighbour atoms lie in the nearest and next-nearest planes. We therefore need to use the extension for distant neighbours given by equation (2.25). First we define a coordinate system such that z lies along [011], y lies along [01$\bar{1}$], and x lies along [100]. The coordinates of the nearest-neighbour atoms are

$$\text{nearest-neighbour:} \quad \frac{a}{\sqrt{2}}\left(\frac{1}{\sqrt{2}},\frac{1}{2},\pm\frac{1}{2}\right) \quad ; \quad \frac{a}{\sqrt{2}}\left(\frac{1}{\sqrt{2}},-\frac{1}{2},\pm\frac{1}{2}\right)$$
$$\text{next-nearest-neighbour:} \quad \frac{a}{\sqrt{2}}(0,0,1) \tag{2.38}$$

We note that this set of neighbours defines two neighbouring planes, so we need to use equation (2.25) for $p = 1, 2$. Moreover, along this direction the two transverse force constants are not identical. However, the zone boundary along this direction is identical to the zone boundary along [001].

The force constants for this direction can readily be calculated using the methods described above for the coordinates (equation (2.38)), giving:

Table 2.2: *Results for the force constant K. Units are J m^{-2}.*

	Ne	Ar	Kr	Xe
[111] L	0.45	1.37	1.54	1.55
[111] T	0.37	1.22	1.35	1.41
[001] L	0.45	1.32	1.54	1.55
[001] T	0.43	1.36	1.48	1.57
mean	0.42 ± 0.04	1.32 ± 0.07	1.48 ± 0.09	1.52 ± 0.07

Table 2.3: *Energy parameters for the rare-gas solids. The experimental values for the lattice energies are taken from the compilation given by Kittel (1976, p 77).*

	Ne	Ar	Kr	Xe
σ (Å)	2.813	3.347	3.563	3.861
ε (kJ mol^{-1})	0.350	1.56	1.98	2.39
W (calc.) (kJ mol^{-1})	−2.10	−9.36	−11.9	−14.3
W (obs.) (kJ mol^{-1})	−1.88	−7.74	−11.2	−16.0

$$\begin{aligned} &\text{nearest-neighbour:} \quad J_{\rm L} = J_{\rm T}^{y} = K \quad ; \quad J_{\rm T}^{x} = 2K \\ &\text{next-nearest-neighbour:} \quad J_{\rm L} = K \quad ; \quad J_{\rm T}^{x} = J_{\rm T}^{y} = 0 \end{aligned} \tag{2.39}$$

where the superscripts on the transverse force constants label the directions of motion. Using the estimates of K given in Table 2.2 the calculated dispersion curves are shown in Figures 2.7–2.10. It can be seen that the agreement is as good as for the other directions.

We finally come back to our main point, namely the evaluation of the parameters in the assumed interatomic potential, equation (2.26). Using the values of the cell parameters given in Table 2.1 and the average values of K given in Table 2.2, we can use equations (2.27) and (2.28) to obtain values for σ and ε, which are given in Table 2.3. We can then calculate the lattice energy W for each system, which is equal to -3ε. The calculated values are compared with experimental data in Table 2.3. The agreement between the calculated and measured lattice energies is reasonably good. It might be expected that the results could be improved by including second-nearest-neighbour interactions, and by fitting the force constants against the dispersion curves at all wave vectors rather than only at the zone boundary points.

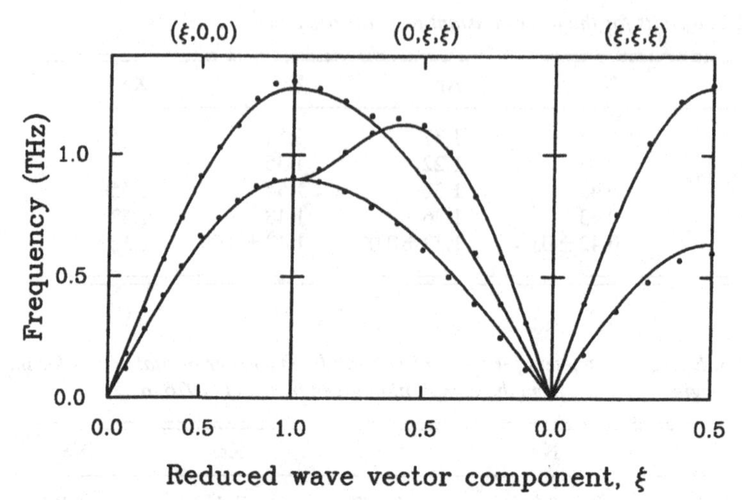

Figure 2.10: Acoustic mode dispersion curves for xenon (Lurie et al. 1974). The lines give the calculated dispersion curves using the model described in the text.

Summary

1. We have defined and justified the use of the harmonic approximation.
2. We have obtained a general expression for the motion of an atom as given by its displacement $u(t)$ – equation (2.5).
3. We have calculated the frequency ω_k for any vibrational mode of wave vector k for our simple model.
4. We have obtained the complete set of allowed values of the wave vector k, and from this we have introduced the concept of the Brillouin zone.
5. All we now need in order to complete the picture is an expression for the mode amplitude $\tilde{u}_k$. This will be dealt with in Chapter 4. Also note that we have not included the relative phase of any mode.
6. We have established the connection between sound waves and the elastic properties – the essential link between microscopic interactions and macroscopic behaviour.
7. We have defined the following general concepts:
   - acoustic modes;
   - longitudinal and transverse modes;
   - dispersion curves.
8. We have extended our ideas from one-dimensional chains to three-dimensional crystals.

FURTHER READING

Ashcroft and Mermin (1976) ch. 22
Cochran (1973) ch. 3
Kittel (1976) ch. 4

3

Dynamics of diatomic crystals: general principles

The methods of the previous chapter are extended to the case of two atoms in the unit cell. The picture that emerges is then generalised for more complex cases, leaving aside the formalism for further consideration in Chapter 6. The lattice dynamics of ionic crystals, covalently bonded crystals, and molecular crystals are discussed.

The basic model

In this chapter we will build upon the results of the previous chapter with the analysis of a more complex model, namely a harmonic chain with two different atom types in the unit cell. The model is shown in Figure 3.1. The force constants G and g, masses M and m, and displacements U and u are defined in this figure. All atoms are separated by $a/2$ when at rest. We will consider only the longitudinal motions, as the extension to include transverse motions is the same as for the monatomic case. We will also only consider nearest-neighbour interactions. This model will give us some new important characteristics, but despite its simplicity it proves to be about as far as we can go before the arithmetic becomes too complicated. For anything more complex we need to resort to the computer, as discussed in Chapter 6.[1]

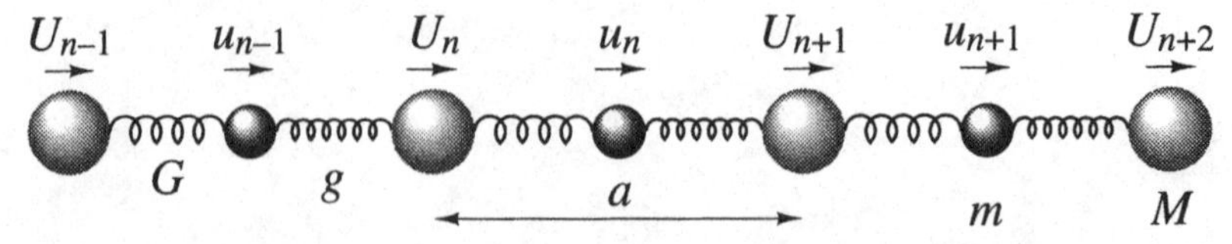

Figure 3.1: Diatomic linear harmonic chain. The parameters are defined in the text.

[1] That said, it is interesting to note that Madelung (1910) undertook an analysis of the lattice dynamics of NaCl even before the atomic structure of crystals had been experimentally verified (Bragg 1913), and one of the major studies of the lattice dynamics of NaCl was performed in 1940 (Kellerman 1940).

Equations of motion

The harmonic energy for our model is given as

$$E = \frac{1}{2}\sum_n \left[G(U_n - u_n)^2 + g(u_{n-1} - U_n)^2 \right] \tag{3.1}$$

Following the procedure we used for the analysis of the monatomic chain in the previous chapter, we can write down the equations of motion for our model:

$$\begin{aligned} M\frac{\partial^2 U_n}{\partial t^2} &= -\frac{\partial E}{\partial U_n} \\ &= -G(U_n - u_n) - g(U_n - u_{n-1}) \\ &= -(G+g)U_n + gu_{n-1} + Gu_n \end{aligned} \tag{3.2}$$

$$\begin{aligned} m\frac{\partial^2 u_n}{\partial t^2} &= -\frac{\partial E}{\partial u_n} \\ &= -g(u_n - U_{n+1}) - G(u_n - U_n) \\ &= -(G+g)u_n + gU_{n+1} + GU_n \end{aligned} \tag{3.3}$$

We can assume the same general solutions as used for the monatomic chain:

$$U_n = \sum_k \tilde{U}_k \exp(i[kna - \omega_k t]) \tag{3.4}$$

$$u_n = \sum_k \tilde{u}_k \exp(i[kna - \omega_k t]) \tag{3.5}$$

where k is the wave vector, ω_k is the angular frequency of a given mode, and $\tilde{U}_k$ and $\tilde{u}_k$ are the two amplitudes for a single given mode (the two atoms each have a different amplitude for any one mode).

There is a point of potential confusion here. The amplitudes are in general complex, and contain the information about the relative phases of the motions of the two atoms. We have defined the wave equations for the two atoms so that they have the same origins, by which we mean that the exponents give the positions of the origins of the unit cells rather than the mean positions of the individual atoms. We could instead have written down the general solutions with the atomic positions, which would mean replacing na by $(n+\frac{1}{2})a$ in one of equations (3.4) or (3.5). The phase difference would then have been transfered to the amplitude, and we would end up with the same final equations.

We perform the same analysis as we used in the previous chapter for the monatomic chain. We substitute the two solutions (3.4) and (3.5) for an individual wave vector into the equations of motion, (3.2) and (3.3). After cancelling the factors common to both sides of the resultant equations we obtain the following simultaneous equations:

$$\begin{aligned} -M\omega_k^2\tilde{U}_k &= -(G+g)\tilde{U}_k + \left(G+g\exp(-ika)\right)\tilde{u}_k \\ -m\omega_k^2\tilde{u}_k &= -(G+g)\tilde{u}_k + \left(G+g\exp(ika)\right)\tilde{U}_k \end{aligned} \tag{3.6}$$

We can write these simultaneous equations in the matrix form:

$$\begin{pmatrix} M\omega_k^2-(G+g) & G+g\exp(-ika) \\ G+g\exp(ika) & m\omega_k^2-(G+g) \end{pmatrix}\begin{pmatrix} \tilde{U}_k \\ \tilde{u}_k \end{pmatrix} = 0 \tag{3.7}$$

For this equation to have a solution, the determinant of the matrix must equal zero. This then gives us a quadratic equation for the square of the angular frequency:

$$\begin{aligned} &\left[M\omega_k^2-(G+g)\right]\left[m\omega_k^2-(G+g)\right] = \left[G+g\exp(ika)\right]\left[G+g\exp(-ika)\right] \\ &\Rightarrow Mm\omega_k^4-(M+m)(G+g)\omega_k^2+4Gg\sin^2(ka/2)=0 \\ &\Rightarrow \omega_k^2 = \frac{(M+m)(G+g)}{2Mm} \pm \frac{\left((M+m)^2(G+g)^2-16MmGg\sin^2(ka/2)\right)^{1/2}}{2Mm} \end{aligned} \tag{3.8}$$

This equation can easily be solved, although the general solution is somewhat cumbersome. The important point to note is that there are two solutions, so that we will have two curves – usually called *branches* – in the dispersion diagram. This follows from the fact that the two atoms have given two equations of motion. If we had also considered the transverse modes (now allowing for three-dimensional motion) we would then have had six equations and six branches. By extension, the number of branches in a three-dimensional crystal with Z atoms in the unit cell is $3Z$.

Solution in the long-wavelength limit

When k is small, we can solve equation (3.8) for the angular frequencies by taking the linear limit on the sine:

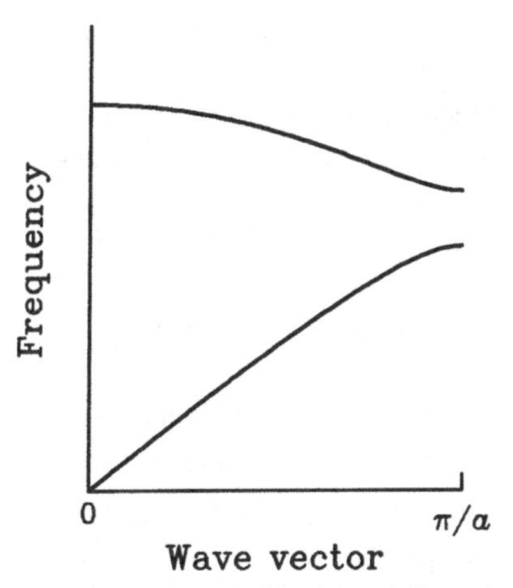

Figure 3.2: Dispersion curves for the one-dimensional diatomic chain.

$$Mm\omega_k^4 - (M+m)(G+g)\omega_k^2 + Ggk^2a^2 = 0$$

$$\Rightarrow \omega_k^2 = \frac{(M+m)(G+g)}{2Mm} \pm \frac{1}{2Mm}\left[(M+m)^2(G+g)^2 - 4MmGgk^2a^2\right]^{1/2}$$

$$= \frac{(M+m)(G+g)}{2Mm}\left[1 \pm \left(1 - \frac{2MmGgk^2a^2}{(M+m)^2(G+g)^2}\right)\right] \tag{3.9}$$

Note that we have taken a series expansion of the square root for the limit $k \to 0$, and have retained only the lowest-order term.

Equation (3.9) has the two roots for the frequency:

$$\omega_k^2 = \frac{(M+m)(G+g)}{Mm} - O(k^2) \quad ; \quad \frac{Ggk^2a^2}{(M+m)(G+g)} \tag{3.10}$$

Clearly the first frequency is large and varies only weakly with k, whilst the second frequency has the same behaviour as the acoustic mode we met in the model for the monatomic chain. The first branch (flat at $k = 0$) is called the *optic mode*,[2] partly because it has a frequency that is in the vicinity of the optical region of the electromagnetic spectrum[3] (whereas the acoustic mode has an acoustic frequency), but also because the atomic motions associated with this branch are the same as the response to an oscillating electromagnetic field. We

[2] These modes are often called *optical modes* in older texts.

[3] Actually in the infrared region of the spectrum, with typical frequency values 1–30 THz.

will discuss this later. For a more complicated crystal with Z atoms in the unit cell, there will always be 3 acoustic branches (one longitudinal and two transverse) and $3(Z-1)$ optic branches.

We expect that the frequencies of the acoustic and optic branches will curve with increasing wave vector, as we expect that $\partial\omega/\partial k = 0$ at $k = \pi/a$. One complete solution to equation (3.8) for all wave vectors is given in Figure 3.2.

Some specific solutions

Solution for equal force constants: representation of a crystal with two different but equally spaced atoms

One simplification to the model is to make the two force constants equal, i.e. $G = g$. This model corresponds to having two types of atoms with constant spacing between them, such as the alkali halides. From equation (3.10) the angular frequencies for small k are

$$\omega_k^2 = 2G\left(\frac{1}{M} + \frac{1}{m}\right) \quad ; \quad \frac{Gk^2a^2}{2(M+m)} \tag{3.11}$$

Inserting these two values into the matrix equation of motion (3.7), we can solve for the relationship between $\tilde{U}_k$ and $\tilde{u}_k$. We find for the acoustic mode in the limit $k = 0$ that both atoms move in the same direction by the same amount:

$$\tilde{U}_0 = \tilde{u}_0 \tag{3.12}$$

This is what we would expect for a vibration that corresponds to compressional motion. On the other hand, for the optic mode we obtain:

$$M\tilde{U}_0 = -m\tilde{u}_0 \tag{3.13}$$

The relative motions are opposite for the two atom types. If they have opposite charges, then their motions correspond to their response to an electric field, which is the point we made concerning the optical character above. The two modes of motion are shown in Figure 3.3.

At the Brillouin zone boundary $k = \pi/a$, and $\sin^2(ka/2) = 1$. In this case equation (3.8) with $G = g$ has the solutions:

$$\omega_k^2 = \frac{2G}{M} \quad ; \quad \frac{2G}{m} \tag{3.14}$$

The matrix equation of motion (3.7) for $k = \pi/a$ simplifies to the form:

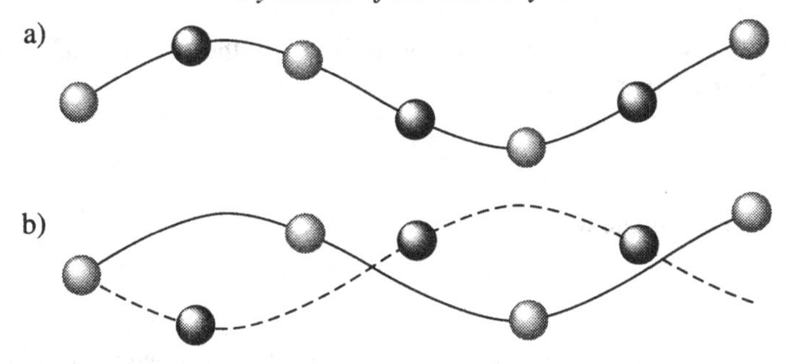

Figure 3.3: Exaggerated transverse atomic motions for *a*) acoustic mode (motions are in phase) *b*) optic mode (motions are out of phase).

$$\begin{pmatrix} M\omega_k^2 - 2G & 0 \\ 0 & m\omega_k^2 - 2G \end{pmatrix} \begin{pmatrix} \tilde{U}_k \\ \tilde{u}_k \end{pmatrix} = 0 \tag{3.15}$$

When we substitute the two angular frequencies from equation (3.14) into this equation, we obtain the solutions for the associated atomic displacements:

$$\omega_k^2 = \frac{2G}{M} \Rightarrow \tilde{u}_k = 0 \ , \ \tilde{U}_k \text{ is undetermined} \tag{3.16}$$

$$\omega_k^2 = \frac{2G}{m} \Rightarrow \tilde{U}_k = 0 \ , \ \tilde{u}_k \text{ is undetermined} \tag{3.17}$$

So for both modes at the zone boundary one atom moves and the other stands still. You should note that this is not a general feature but is specific to this case. However, it is often the case that the distinction between the acoustic and optic modes is lost for wave vectors close to the zone boundary.

Solution for equal masses: representation of a crystal with two atoms of the same type in a primitive unit cell

Another simplification is to set the two masses to equal values, i.e. $M = m$, whilst the two force constants remain unequal. This model could represent the case where there is only one atom type but two distinct environments, so that there are different separation distances, such as in the diamond or hexagonal close-packed structures. The angular frequencies for small k from equation (3.10) become:

$$\omega_k^2 = \frac{2(G+g)}{M} \ ; \ \frac{Ggk^2a^2}{2M(G+g)} \tag{3.18}$$

Inserting these values into equation (3.7) gives the two sets of displacements. For the acoustic mode we have

$$\tilde{U}_0 = \tilde{u}_0 \tag{3.19}$$

and for the optic mode we have

$$\tilde{U}_0 = -\tilde{u}_0 \tag{3.20}$$

These relations are similar to the model for unequal masses but identical force constants.

At the Brillouin zone boundary the solutions of equation (3.8) for $M = m$ are

$$\omega_k^2 = \frac{2g}{M}\ ;\ \frac{2G}{M} \tag{3.21}$$

For $k = \pi/a$ equation (3.7) reduces to

$$\begin{pmatrix} M\omega_k^2 - (G+g) & (G-g) \\ (G-g) & M\omega_k^2 - (G+g) \end{pmatrix} \begin{pmatrix} \tilde{U}_k \\ \tilde{u}_k \end{pmatrix} = 0 \tag{3.22}$$

Substitution of the angular frequencies from equation (3.21) gives the associated solution for the atomic displacements:

$$\omega_k^2 = \frac{2g}{M} \Rightarrow \tilde{U}_k = \tilde{u}_k \tag{3.23}$$

$$\omega_k^2 = \frac{2G}{M} \Rightarrow \tilde{U}_k = -\tilde{u}_k \tag{3.24}$$

These results contrast with the case for unequal masses and identical force constants. Our model now retains the distinction between the acoustic and optic modes for all wave vectors.

Solution for equal masses but with $G \gg g$: representation of a simple molecular crystal

The case when the masses are equal but the force constants are very different is representative of systems where there are both strong covalent bonds and weak interactions. One simple example is a lattice of diatomic molecules (e.g. molecular nitrogen), where the two atoms are tightly bound, and the molecules interact via weak dispersive and repulsive interactions. Anticipating our results, we would expect the molecules to move as rigid bodies, with the internal molecular vibrations being separated from the normal lattice dynamics. Another example is any mineral containing SiO_4 tetrahedra. In this case the Si–O bond stretching vibrations are of much higher frequency than the vibrations which involve displacements, rotations and bond-bending distortions of the tetrahedra.

We can now re-write our general equation (3.8) for our limiting case $G \gg g$ as

$$Mm\omega_k^4 - (M+m)G\omega_k^2 + 4Gg\sin^2(ka/2) = 0 \tag{3.25}$$

The roots of the quadratic equation are given as:

$$\omega_k^2 = \frac{4g}{(M+m)}\sin^2(ka/2)\ ;\ \frac{G(M+m)}{Mm} \tag{3.26}$$

with the two solutions for all k in the limiting case given as (respectively):

$$\tilde{U}_k = \tilde{u}_k \tag{3.27}$$

$$M\tilde{U}_k = -m\tilde{u}_k \tag{3.28}$$

The first solution represents an acoustic mode in which the molecule moves as a rigid body with just one degree of freedom. This solution is equivalent to the solution for the monatomic chain, equation (2.7), with the atomic mass replaced by the molecular mass. The second solution, on the other hand, represents the internal molecular bond-stretching frequency, which is effectively constant for all wave vectors.

Generalisation to more complex cases

Acoustic and optic modes

We reinforce our descriptions of acoustic and optic modes by considering the case of four atoms in a unit cell, all moving along the x direction, and with $k \approx 0$. The atomic displacements associated with the four modes of vibration are given in Table 3.1. These are the only four independent linear combinations of positive or negative displacements. Except for mode 1, the sums of the displacements are zero. Thus modes 2 to 4 correspond to the optic modes. Mode 1, for which all the atoms move in phase, is the acoustic mode. All other possible displacement patterns correspond to linear combinations of these basic modes. For example:

Table 3.1: *Atomic displacements for four modes.*

	Atom 1	Atom 2	Atom 3	Atom 4
Mode 1	+1	+1	+1	+1
Mode 2	+1	−1	+1	−1
Mode 3	+1	+1	−1	−1
Mode 4	+1	−1	−1	+1

$$\begin{array}{lrrrrcl} \text{displacements} & +1 & -1 & -1 & -1 & = & \tfrac{1}{2}\times \text{modes } (2+3+4-1) \\ \text{displacements} & +1 & 0 & 0 & 0 & = & \tfrac{1}{4}\times \text{modes } (1+2+3+4) \\ \text{displacements} & +3 & -1 & -1 & -1 & = & \text{modes } (2+3+4) \end{array}$$

It is easy to generalise this argument for Z atoms per cell, showing that for displacements along one direction there will always be 1 acoustic mode and $(Z-1)$ optic modes. If we now allow motions along the y and z directions as well, it is clear that we will have 3 acoustic modes and $(3Z-3)$ optic modes. We can conclude that the principal difference between acoustic and optic modes is that in the limit $k \to 0$ the acoustic modes have all the atoms moving in phase, whereas the optic modes have the atoms moving out of phase. This distinction is not relevant for larger wave vectors, as we have seen in one case for our simple one-dimensional model. Note that if the atoms have different masses, the displacement patterns described above will be for mass-weighted displacements.

Zone boundary behaviour

We pointed out earlier that the gradient of the dispersion curve goes to zero at the boundaries of the Brillouin zone. This feature was noted for high symmetry cases. In the general case, when the direction of the wave vector is not normal to the plane of the surface of the Brillouin zone, it is the normal component that has a zero gradient at the zone boundary.

There is one significant exception to this. If the crystal space group has a screw-axis or a glide plane,[4] the frequencies of pairs of branches for some zone boundary wave vectors become degenerate, with the gradient of the upper branch equal to the negative of the gradient of the lower branch. This feature is illustrated in some of the examples given below. The simplest explanation of this behaviour is that wave vectors along screw-axes are completely insensi-

[4] Such a crystal is often called *non-symmorphic*.

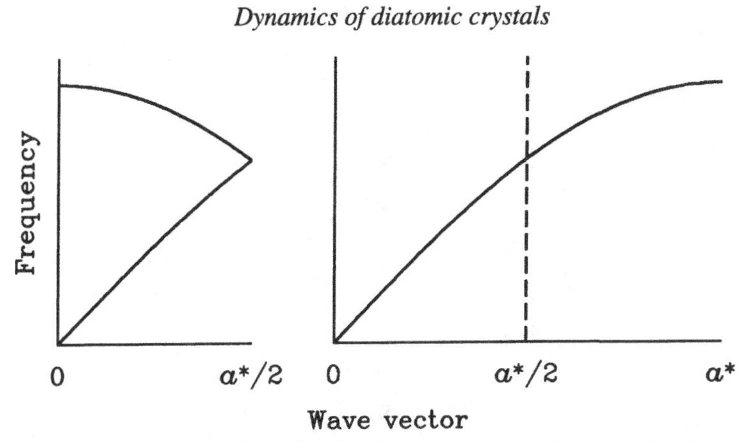

Figure 3.4: The behaviour of the dispersion curve at a zone boundary along the direction of a screw-axis. The dispersion curve on the left hand side can be *unfolded* as on the right hand side, to give the dispersion curves for two symmetries.

tive to the rotations of the screw-axis, and hence it will appear as if the unit cell has only half the repeat distance. This being so, the Brillouin zone will appear to be twice as large, and the real zone boundary will not be noticed. It is then possible to consider the dispersion curves for wave vectors along these special directions in an extended zone scheme, as illustrated in Figure 3.4.

Symmetry of lattice vibrations

Any lattice vibration will lower the symmetry of the structure. Different modes will have different effects on the symmetry, and it is possible to assign a symmetry to each mode, which labels the symmetry operations that are not removed by the vibration. In the language of group theory, each vibrational mode will correspond to a single irreducible representation of the point group of the particular wave vector.

The main effect of symmetry is that vibrations of the same symmetry can interact, whereas those of different symmetry cannot. What this means is that the displacement patterns of two modes of the same symmetry can mix, so that one cannot identify a unique displacement pattern to a mode if another mode of the same symmetry exists. Related to this is the fact that two modes of the same symmetry cannot have dispersion curves that cross. Instead of crossing, two modes will appear to cross but will in fact "repel" each other as they approach each other in frequency. This effect is known as *anti-crossing*, and is illustrated in Figure 3.5. Anti-crossing modes are quite common, and will be seen in the examples shown below. Symmetry is described in more detail by Bradley and Cracknell (1972).

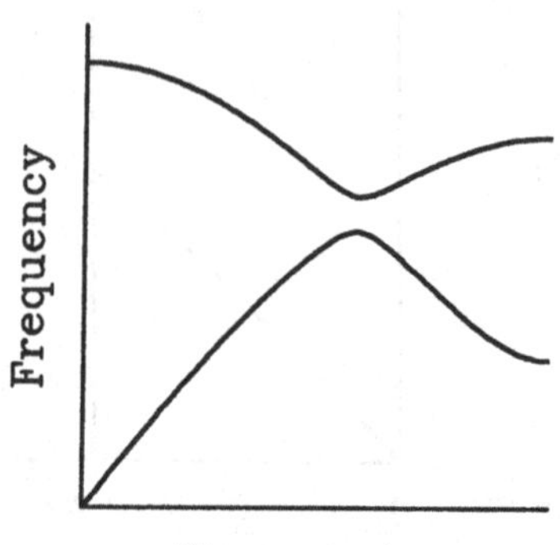

Figure 3.5: Representation of anti-crossing in a dispersion diagram.

Examples

Dispersion curves have now been measured in a wide variety of materials using inelastic neutron scattering; here we give some representative examples. It should be appreciated that the determination of a complete set of dispersion curves for anything other than the simplest structures is a major undertaking, and frequently only an incomplete set of branches is obtained. The usual motivation for measuring dispersion curves is the information they give about interatomic forces, and often an incomplete set of dispersion curves is adequate for this. We will stress the importance of the particular examples in this regard. As we will see in Chapter 9, the low-frequency branches are more easily measured by neutron scattering than the high-frequency modes. But the low-frequency modes have the greater dispersion and are therefore the most interesting. The high-frequency modes are often only weakly dependent on wave vector, and therefore a spectroscopic measurement (Chapter 10) at $\mathbf{k} \approx 0$ will generally suffice for most practical purposes. One of the key points to note from the examples given here is the range of values of frequencies of the lattice vibrations, typically in the THz region. Another feature that will be apparent is the complexity of the dispersion curves! A catalogue of measured dispersion curves in insulators is given in Bilz and Kress (1979), and an update of this compilation is given in Appendix H.

Ionic crystals: the alkali halides

The alkali halides are the best-studied family of insulator crystals, primarily because of their relative simplicity and the ease with which high-quality single crystals could be grown for neutron scattering experiments. Dispersion curves for NaCl and KBr are shown in Figure 3.6.[5] There are two ions in the primitive

[5] Dispersion curves for all the alkali halides are given in Bilz and Kress (1979).

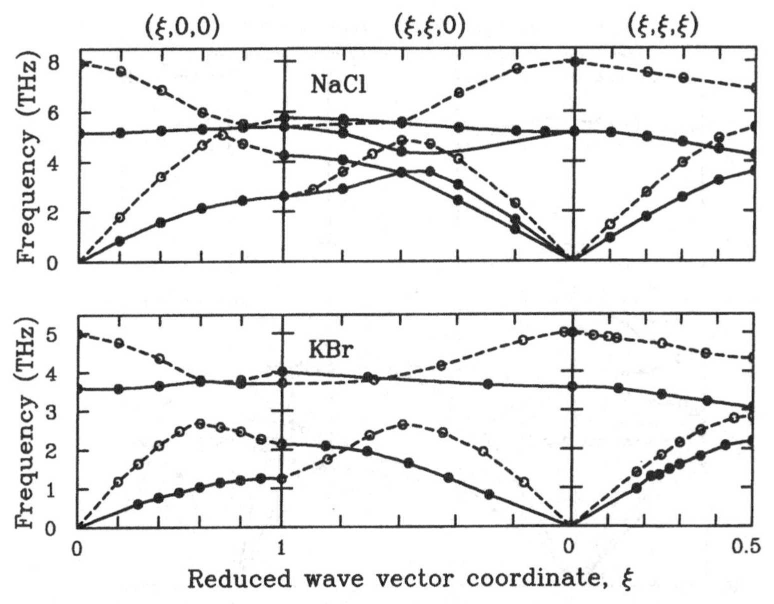

Figure 3.6: Measured dispersion curves for NaCl (Raunio et al. 1969) and KBr (Woods et al. 1963). The transverse modes are shown as filled circles and continuous curves, and the longitudinal modes are shown as open circles and dashed curves. The data for KBr along [110] are incomplete.

cell, giving 3 acoustic and 3 optic modes for all wave vectors. Note, however, the degeneracies of the transverse modes, acoustic and optic, for wave vectors along the [001] and [111] symmetry directions. Also note that some branches show significant anti-crossing effects.

The dispersion curves for the alkali halides highlight one feature of ionic crystals, namely that the longitudinal and transverse optic modes have different frequencies at the zone centre. At first sight this is a surprising observation, since one might naively expect that transverse and longitudinal optic modes should have the same frequency. The difference arises from the fact that the longitudinal and transverse modes give rise to different long-range fields in the limit $\mathbf{k} \to 0$. This difference in frequencies is colloquially called the *LO/TO splitting*. For the alkali halides this observation is described by the Lyddane–Sachs–Teller (LST) relation (Lyddane et al. 1941):

$$\frac{\varepsilon_0}{\varepsilon_\infty} = \frac{\omega_{\mathrm{L}}^2}{\omega_{\mathrm{T}}^2} \tag{3.29}$$

where ε_0 and ε_∞ are the dielectric constants for oscillating electric fields at zero and infinite frequency respectively, and ω_{L} and ω_{T} are the longitudinal and

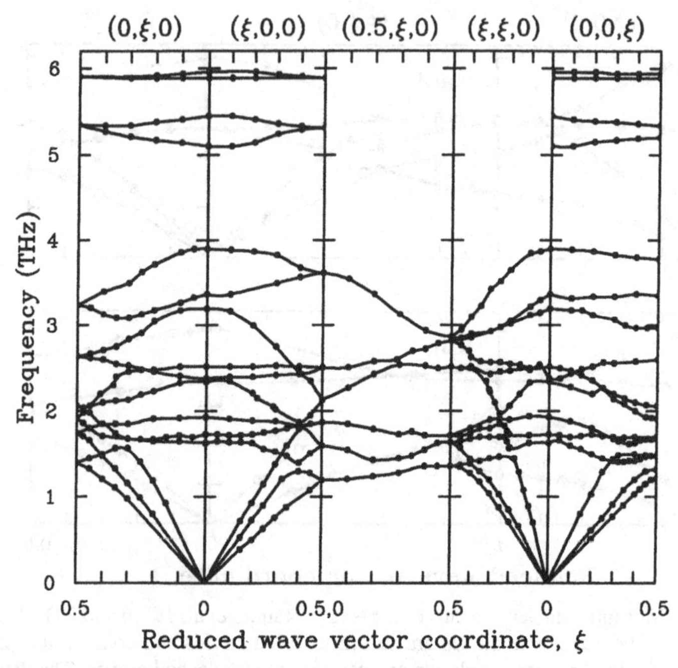

Figure 3.7: Measured dispersion curves for the molecular crystal naphthalene (Natkaniec et al. 1980).

transverse optic mode frequencies in the limit $\mathbf{k} \rightarrow 0$. The LST relation can be generalised for more complex crystal structures (Cochran 1959b; Cochran and Cowley 1962). The LST relation is derived in many solid state textbooks using standard dielectric theory which is outside the scope of the present book (Ashcroft and Mermin 1976, p 548; Kittel 1976, p 306). We will meet the LST relation again in connection with the origin of displacive phase transitions (Chapter 8).

There was a burst of activity in the 1960s, when neutron scattering became a routine tool, to measure the dispersion curves for most alkali halide crystals. The motivation for this work was the development of force constant models that could be tested against experimental data. During this activity the shell model was found to be essential to model the lattice dynamics of these systems (Woods et al. 1960, 1963; Cochran 1971), and this work led to many refinements of the basic model (Bilz and Kress 1979, pp 8–14). The significance of these developments has been in the insights they have provided into crystal stability, which is a particularly crucial issue in the study of displacive phase transitions (Cochran 1959c, 1960, 1961). The soft mode model for ferroelectric phase transitions, described in Chapter 8, arose from this work.

Apart from the alkali halides, the dispersion curves in a number of related systems has also been measured, e.g. MgO (Sangster et al. 1970). More complex ionic crystals, such as CaF_2 (Elcombe and Pryor 1970), TiO_2 (Traylor et al. 1971) and perovskite $SrTiO_3$ (Stirling 1972) have also been studied by inelastic neutron scattering. The latter system was studied in order to provide information about the forces that give rise to the displacive phase transitions as a detailed case study.

Molecular crystals: naphthalene

One of the main features of molecular crystals as far as the lattice dynamics are concerned is that there is usually a large difference between the frequencies of modes in which the molecules move as rigid units (the *external modes*) and the modes which involve distortions of the molecules (the *internal modes*). Typically the external modes have frequencies below 3–4 THz, whereas the highest internal mode frequencies, which will involve stretching of covalent bonds, will be an order of magnitude higher. The internal modes will usually have only a weak dependence on wave vector, as we found earlier in the example with very different force constants.

The external modes involve translations and rotations of the molecules, giving six degrees of freedom per molecule. For Z molecules in the unit cell there will be 3 acoustic modes and $(6Z - 3)$ external optic modes. If there are f atoms per molecule, there will be a total of $(3Zf - 3)$ optic modes, of which $3Z(f - 2)$ will be the internal modes.

One system that has been studied in considerable detail is naphthalene, $C_{10}H_8$, which has two molecules in the unit cell. Neutron scattering studies of the deuterated form have led to the identification of all 12 external branches for the dispersion curves with wave vectors along the major symmetry directions in reciprocal space (Natkaniec et al. 1980). The lowest-frequency internal modes have also been measured. The results are shown in Figure 3.7. The space group of naphthalene is monoclinic $P2_1/c$, with the 2_1 screw-axis along [010]. The dispersion curves at $(0, \frac{1}{2}, 0)$ show the degeneracies due to the screw-axis that was discussed above, and additional degeneracies are found for the zone boundary wave vectors between $(\frac{1}{2}, 0, 0)$ and $(\frac{1}{2}, \frac{1}{2}, 0)$. The dispersion curves also show considerable anti-crossing effects. The motivation for this work was to test basic models for interatomic interactions for hydrocarbons. A basic Buckingham model was found to work fairly well, but it was later shown that the calculations of the dispersion curves could be improved by including a quadrupole moment on the molecules (Righini et al. 1980). Extensive work was also performed on the related material anthracene ($C_{14}H_{10}$), which was

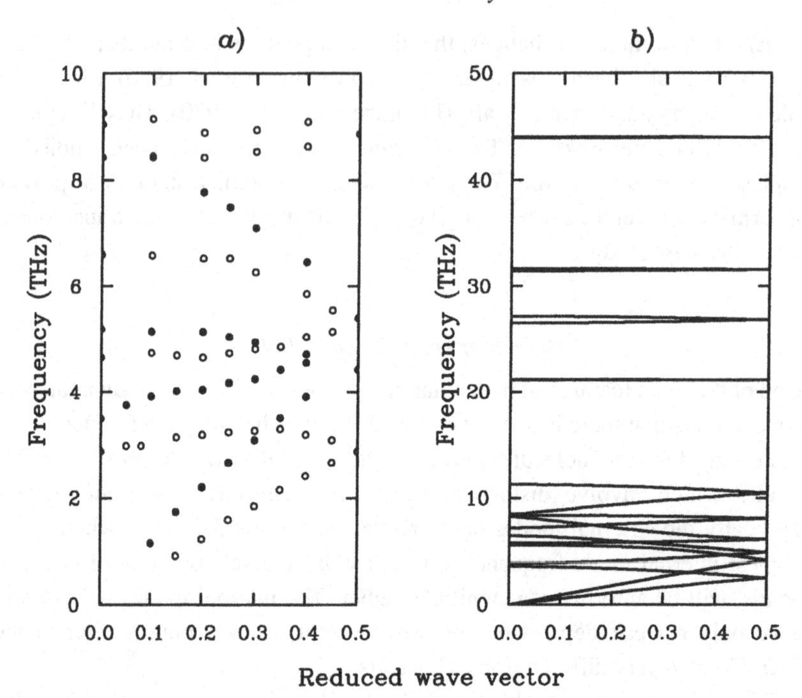

Figure 3.8: *a*) Measured dispersion curves of calcite for wave vectors along the three-fold axis (Cowley and Pant 1973). *b*) Calculated dispersion curves along the same axis (Dove et al. 1992c).

chosen to illustrate the relationship between the external and lowest-frequency internal modes (Dorner et al. 1982). Other molecular systems that have been studied in some detail include hexamethylene-tetramine (Dolling and Powell 1970), $C_4N_2H_4$ (Reynolds 1973), $C_2(CN)_4$ (Chaplot et al. 1983), and $C_6F_3Cl_3$ (Dove et al. 1989). These studies have provided useful insights into the cohesive energies of molecular crystals.

Ionic molecular crystals: calcite

Calcite, $CaCO_3$, is an example of a crystal with a number of distinctly different interactions. The atoms in the CO_3 groups are held together by strong covalent bonds, so that the carbonate groups move as rigid bodies in the low-frequency modes. On the other hand, there are strong Coulombic interactions between the Ca cations and the carbonate groups. Calcite has a rhombohedral structure with two formula units per unit cell; the carbonate groups are oriented normal to the three-fold axis. We therefore expect 18 external modes. Most of the external modes have been measured for wave vectors along the three-fold symmetry

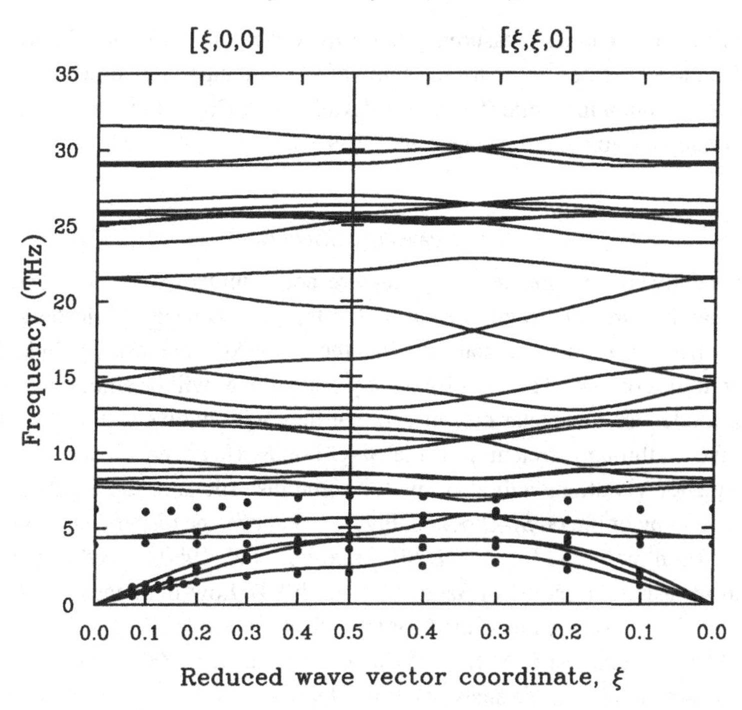

Figure 3.9: Measured and calculated dispersion curves for quartz.

axis; along this direction the transverse modes are doubly-degenerate, thereby reducing the number of branches in the dispersion curves to 12. The experimental results for the 10 lowest branches are shown in Figure 3.8a (Cowley and Pant 1973). Note the degeneracies at the zone boundary that arise from the existence of the glide plane, and the anti-crossing effects.

A model interatomic potential has been developed for calcite, which includes interactions for the external and internal modes, and which was optimised by fitting to the structure, elastic constants, and frequencies at $\mathbf{k} = 0$ (Dove et al. 1992c). The complete set of calculated dispersion curves for wave vectors along the same direction as the experimental data is shown in Figure 3.8b. The agreement with the experimental data is about as good as can be obtained using models of the form described in Chapter 1. Better agreement can usually be obtained with the use of force constant models, but this may be due to the fact that such models can have more free parameters and fewer constraints. One feature that should be noted from Figure 3.8b is the separation between the external and internal mode frequencies. The external mode frequencies are all below 12 THz, whereas the internal mode frequencies are in the region 20–44 THz. The example of calcite provides an illustration of the value of lattice dynamics measurements for the development of model potentials. The model has been used for the study of a high-temperature order–disorder phase transition in calcite (Dove and Powell 1989; Dove et al. 1992c), and for mineralogical applications (Dove et al. 1992c).

Covalent crystals: silicates

The covalently bonded silicates are in many respects similar to molecular crystals, in that they are composed of rigid SiO_4 tetrahedra. However, silicates are different from molecular crystals in that the tetrahedra are usually linked together at the corners.[6] The high-frequency optic modes will be similar to the internal modes for molecular crystals, in that they will involve distortions of the tetrahedra through stretching of the Si–O bonds. However, modes of the low-frequency vibrations will also involve some distortions of the tetrahedra through bending of the O–Si–O bonds. Figure 3.9 shows the dispersion curves that have been measured for quartz, SiO_2 (Dorner et al. 1980), together with calculations using the model of Sanders et al. (1984). Low-frequency dispersion curves have also been measured for the silicates forsterite, Mg_2SiO_4 (Rao et al. 1988), and andalusite, Al_2SiO_5 (Winkler and Buehrer 1990). These measurements are of value in the analysis of transferable model interatomic potentials for silicates.

High-temperature superconductors

Classical superconductivity in metals involves an interaction between pairs of electrons mediated by the lattice vibrations. The record high temperature for the existence of superconductivity has recently risen dramatically with the discovery of superconductivity in oxides containing copper that have structures related to the perovskite structure. The crystal chemistry of these so-called *high-temperature superconductors* is not simple, as the superconductivity is associated with depletion of the oxygen content accompanied by changes in the valence of the copper ions. The two classic examples are La_2CuO_{4-x}, with superconductivity found below 45 K, and $YBa_2Cu_3O_{7-x}$, with superconductivity found below 96 K. It is not clear at the time of writing whether the superconductivity is associated with the lattice dynamics in the same way as it is in metals, or whether the superconducting mechanism involves pairing of electrons by another mechanism. However, the lattice dynamics are certainly involved in one way, in that the basic crystal structures are susceptible to dis-

[6] There are silicates that have unconnected SiO_4 tetrahedra.

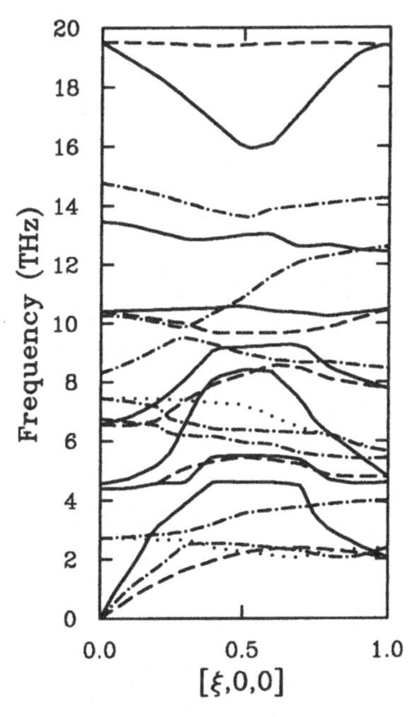

Figure 3.10: Dispersion curves along one symmetry direction in the high-temperature superconductor La_2NiO_4 (Pintschovius et al. 1991). The different types of lines represent different symmetries.

placive phase transitions depending on the oxygen content, and the structural state does have an effect on the existence of the superconducting state. We will see in Chapter 8 that there is a close connection between displacive phase transitions and lattice dynamics. For all these reasons it is desirable to measure the dispersion curves in high-temperature superconductors and related materials. A number of high-temperature superconductor samples have had their lattice dynamics measured (Birgeneau et al. 1987; Böni et al. 1988; Pintschovius et al. 1989, 1991; Reichardt et al. 1989). An example that illustrates the relative complexity of the dispersion curves in these materials is shown in Figure 3.10.

Summary

1 For Z atoms in the unit cell there are $3Z$ branches in the dispersion diagram, corresponding to $3Z$ modes of motion.
2 There are always 3 acoustic branches for any wave vector, with all atoms moving in phase for small wave vectors.
3 The remaining branches are optic modes, with atoms moving out of phase. Optic mode frequencies remain non-zero for zero wave vector.

FURTHER READING

Ashcroft and Mermin (1976) ch. 22
Bilz and Kress (1979)
Califano et al. (1981)
Cochran (1973) ch. 3, 7
Kittel (1976) ch. 4

4

How far do the atoms move?

In this chapter we tackle the question of the amplitudes of lattice vibrations. We start by defining the normal mode coordinate. The effect of quantum mechanics on the lattice vibrations introduces temperature into the theory.

So far we have introduced most of the important features of lattice dynamics, namely the dispersion curves, acoustic and optic branches, the harmonic approximation, allowed wave vectors etc. We have used only simple models, but the features these have given are general and are found in complex cases as well. Leaving aside for now the discussion of the formal methods used for the calculation of dispersion curves and displacement vectors in general cases (dealt with in Chapter 6), we next need to tackle the question of the actual amplitudes of the lattice vibrations. We will need to introduce some formalism – the concept of normal modes and normal mode coordinates – in order to be able to draw some general conclusions.

Normal modes and normal mode coordinates

In Chapter 1 we introduced the general wave equation for a three-dimensional crystal with a monatomic unit cell (equation (1.26)). The main point was that a general wave causes the atoms to move in directions that are not necessarily parallel or perpendicular to the direction of the wave vector. So far we have included the information about the vibrational direction and the amplitude in a single amplitude vector. This means that we would write the equation for the displacement of any atom (labelled as the j-th atom in the l-th unit cell) as

$$\mathbf{u}(jl,t)=\sum_{\mathbf{k},\nu}\mathbf{U}(j,\ \mathbf{k},\nu)\exp\left(i\left[\mathbf{k}\cdot\mathbf{r}(jl)-\omega(\mathbf{k},\nu)t\right]\right) \tag{4.1}$$

where we have summed over all wave vectors, $\mathbf{k}$, and over all branches in the dispersion diagram, ν. Strictly speaking, we should really say that ν labels the mode for any wave vector. $\mathbf{U}(j, \mathbf{k}, \nu)$ is the amplitude vector that tells us how atom j moves under the influence of the wave $(\mathbf{k}, \nu)$, giving the direction and amplitude of the motion.

It is usual to re-write equation (4.1) in a new form:

$$\mathbf{u}(jl,t) = \frac{1}{\left(Nm_j\right)^{1/2}} \sum_{\mathbf{k},\nu} \mathbf{e}(j,\mathbf{k},\nu)\exp\left(i\mathbf{k}\cdot\mathbf{r}(jl)\right)Q(\mathbf{k},\nu) \tag{4.2}$$

where m_j is the mass of the j-th atom and N is the number of unit cells in the crystal. The new quantity $Q(\mathbf{k}, \nu)$ has subsumed the time dependence, and is a complex scalar quantity. In order that the net atomic displacement is always a real quantity, $Q(\mathbf{k}, \nu)$ is subject to the constraint:

$$Q(-\mathbf{k},\nu) = Q^*(\mathbf{k},\nu) \tag{4.3}$$

$Q(\mathbf{k}, \nu)$ gives both the amplitude of the wave and the time dependence. On the other hand, the vector $\mathbf{e}(j, \mathbf{k}, \nu)$, which is parallel to $\mathbf{U}(j, \mathbf{k}, \nu)$, gives the direction in which each atom moves, and is normalised such that

$$\sum_j \left|\mathbf{e}(j,\mathbf{k},\nu)\right|^2 = 1 \tag{4.4}$$

The vector $\mathbf{e}(j, \mathbf{k}, \nu)$ is called the *displacement vector*. Other terms used are the *mode eigenvector*, since it is obtained as a solution of the eigenequations introduced in Chapter 6, or the *polarisation vector*, since it is related to the polarisation of the wave. It should be noted that within the harmonic approximation the displacement vector is independent of the mode amplitude.

It is straightforward to obtain the reverse Fourier transform of equation (4.1):

$$Q(\mathbf{k},\nu) = \frac{1}{N^{1/2}} \sum_{jl} m_j^{1/2} \exp\left(-i\mathbf{k}\cdot\mathbf{r}(jl)\right)\mathbf{e}^*(j,\mathbf{k},\nu)\cdot\mathbf{u}(jl,\,t) \tag{4.5}$$

The vibrational modes that we have calculated are called the *normal modes* of the system. They are travelling waves, each with a unique frequency, and have been defined by the formalism such that each normal mode is *orthogonal*. This is expressed mathematically for two modes labelled ν and ν' as:

$$\sum_j \mathbf{e}(j,\mathbf{k},\nu)\cdot\mathbf{e}(j,-\mathbf{k},\nu') = \delta_{\nu,\nu'} \tag{4.6}$$

where the use of both **k** and –**k** takes account of the imaginary components of **e**.

With the use of the new parameter $Q(\mathbf{k}, \nu)$ the equations for the energy of a system gain a new simplicity. We will see later in Chapter 6 that the dynamic energy of the harmonic system, called the *Hamiltonian* $\mathcal{H}$, can be written as:

$$\mathcal{H} = \frac{1}{2}\sum_{\mathbf{k},\nu} \dot{Q}(\mathbf{k},\nu)\dot{Q}(-\mathbf{k},\nu) + \frac{1}{2}\sum_{\mathbf{k},\nu} \omega^2(\mathbf{k},\nu)Q(\mathbf{k},\nu)Q(-\mathbf{k},\nu) \qquad (4.7)$$

The simplicity of equation (4.7) makes the use of the new variables $Q(\mathbf{k}, \nu)$ so popular. These variables look like coordinates, and so they are given the name of *normal mode coordinates*. An important feature of equation (4.7) is that there are no terms with interactions between different modes or between modes of different wave vectors. This is consistent with one of the features of harmonic systems, that there is no exchange of energy between one vibrational mode and another – each mode is independent of all others. Equation (4.7) represents a good starting point for further development, and we will meet this again in our discussion of anharmonic effects in Chapter 8, and we will also use it in Chapter 11 for detailed consideration of quantum mechanics.

The quantisation of normal modes

Everything that we have done so far has been classical. However, the behaviour of nature is determined by quantum mechanics rather than classical mechanics. For now we will simply cite the effects of quantum mechanics, although we will use some of these results quite a lot later on; the quantum picture will be derived in Chapter 11. Just as light is a wave motion that can be considered as composed of particles called *photons*, we can think of our normal modes of vibration in a solid as being particle-like. We call these little packets of energy *phonons* by analogy, the root of this word meaning *sound*. So we can think of our vibrational modes as being particles, and from now on we will regularly interchange between the particle and wave concepts, and will always call our vibrations phonons. This is the common usage!

Just like light, the energy of a phonon is given by the product of Planck's constant $\hbar$ (= $h/2\pi$) and the angular frequency ω. However, there is one feature that is peculiar to quantum mechanics that we need to take into account, namely that the lowest energy state of a quantum harmonic oscillator E_0 is not zero energy but equal to

$$E_0 = \frac{1}{2}\hbar\omega \qquad (4.8)$$

This is called the *zero-point energy*, and the corresponding motions at $T = 0$ K are called the zero-point vibrations.

The energy of the vibration can be changed only by integral units of the Planck energy, $\hbar\omega$. The mean energy of each vibrational mode, $E(\mathbf{k}, \nu)$, is then given as:

$$E(\mathbf{k}, \nu) = \hbar\omega(\mathbf{k}, \nu)\left[\frac{1}{2} + n(\mathbf{k}, \nu)\right] \tag{4.9}$$

where $n(\mathbf{k}, \nu)$ is the number of phonons in the ν-th branch with wave vector $\mathbf{k}$, and is called the *phonon number* or the *occupation number*. The phonon number is related to the temperature T by the expression

$$n(\mathbf{k}, \nu) = n(\omega, T) = \left[\exp\left(\hbar\omega(\mathbf{k}, \nu) / k_B T\right) - 1\right]^{-1} \tag{4.10}$$

where k_B is *Boltzmann's constant.* This expression is derived in Appendix C, and the function $n(\omega, T)$ is shown in Figure 4.1. It is a special case of the *Bose–Einstein distribution*, and is essentially an energy distribution function for quantum-mechanical particles (in our case the phonons) that are not subject to the Pauli exclusion principle (unlike electrons). It should be noted that the number of phonons depends only on frequency and temperature, and is independent of wave vector and mode label other than via the dependence of frequency on these quantities. This is why we write the phonon number as $n(\omega, T)$ in equation (4.10), to reflect the actual dependence.

Vibrational energies and normal mode amplitudes

We are now in a position to be able to calculate the amplitudes of the normal modes. We first calculate the average kinetic energy, $\langle K \rangle$, of the crystal, defined in the normal way from the atomic velocities:

$$\langle K \rangle = \frac{1}{2}\sum_{jl} m_j \left\langle \left|\dot{\mathbf{u}}(jl,t)\right|^2 \right\rangle \tag{4.11}$$

The instantaneous velocity of a single atom simply follows from equation (4.2) as

$$\dot{\mathbf{u}}(jl,t) = \frac{-i}{\left(Nm_j\right)^{1/2}} \sum_{\mathbf{k},\nu} \omega(\mathbf{k}, \nu)\mathbf{e}(j,\mathbf{k}, \nu)\exp\left(i\mathbf{k} \cdot \mathbf{r}(jl)\right)Q(\mathbf{k}, \nu) \tag{4.12}$$

When we insert this equation for the velocity of an atom into equation (4.11) for the average kinetic energy, the final result ends up looking rather simple:

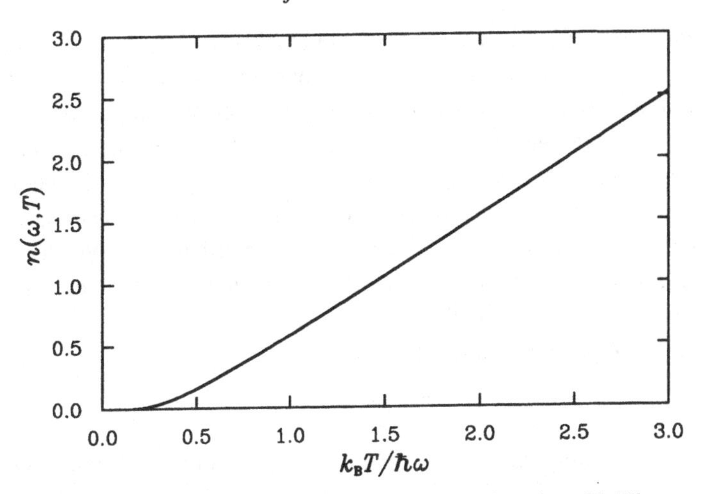

Figure 4.1: The Bose–Einstein distribution $n(\omega, T)$ as a function of $k_BT/\hbar\omega$.

$$\langle K\rangle = \frac{1}{2}\sum_{\mathbf{k},\nu}\omega^2(\mathbf{k},\nu)\left\langle |Q(\mathbf{k},\nu)|^2\right\rangle \tag{4.13}$$

The detailed derivation of this result is given in Appendix B.

We could repeat the whole procedure for the harmonic potential energy of the crystal, V, which is dependent on the squared modulus of the displacement. However, we are not yet in a position to do this, as the calculation requires some of the formalism to be developed in Chapter 6. Instead we use the property of a harmonic oscillator that $\langle V\rangle = \langle K\rangle$. Therefore the total harmonic energy of the crystal is

$$\langle E\rangle = \langle K\rangle + \langle V\rangle = \sum_{\mathbf{k},\nu}\omega^2(\mathbf{k},\nu)\left\langle |Q(\mathbf{k},\nu)|^2\right\rangle \tag{4.14}$$

This is the sum over the energies of the separate modes, so that we are able to conclude that the energy of a single mode, $E(\mathbf{k}, \nu)$, is equal to:

$$E(\mathbf{k},\nu) = \omega^2(\mathbf{k},\nu)\left\langle |Q(\mathbf{k},\nu)|^2\right\rangle \tag{4.15}$$

We are now in a position to calculate the normal mode amplitude. It is simply obtained by combining equations (4.9) and (4.15):

$$\left\langle |Q(\mathbf{k},\nu)|^2\right\rangle = \frac{\hbar}{\omega(\mathbf{k},\nu)}\left(n(\omega,T) + \frac{1}{2}\right) \tag{4.16}$$

We should comment on the thermal average brackets that we have introduced into the picture. Although in an isolated harmonic oscillator there is no exchange of energy between different modes, in practice there is always contact with the outside world to change things around a bit inside the solid. In addition there will in practice always be some exchange of energy between modes because of phonon–phonon collisions. These arise from the breakdown of the harmonic approximation, which we will consider in more detail in Chapter 8. It is sufficient to note though that even allowing for the breakdown of the main model, the central ideas remain intact!

In the high-temperature limit, defined as

$$k_{\mathrm{B}}T \gg \hbar\omega \tag{4.17}$$

the distribution function for the phonon number in equation (4.10) reduces to

$$n(\omega,T)+\frac{1}{2}\cong\frac{k_{\mathrm{B}}T}{\hbar\omega}\gg 1 \tag{4.18}$$

In this case the normal mode amplitude has the simpler form:

$$\left\langle \left|Q(\mathbf{k},\nu)\right|^2\right\rangle = k_{\mathrm{B}}T\,/\,\omega^2(\mathbf{k},\nu) \tag{4.19}$$

At high temperatures, therefore, all modes have the same energy, equal to $k_{\mathrm{B}}T$. We noted at equation (4.14) that half of this energy is kinetic, and half is potential. Thus the kinetic energy per mode in the classical limit is $\frac{1}{2}k_{\mathrm{B}}T$. Since the number of modes is equal to the number of degrees of freedom (3 × *number of atoms*), it is clear that we have also ended up with the classical equipartition result that each degree of motion has an average kinetic energy of $\frac{1}{2}k_{\mathrm{B}}T$.

By combining the results expressed in equations (4.2) and (4.16) we have finally achieved our original goal of obtaining a complete description of the atomic motions in a crystal within the harmonic approximation.

So how far do the atoms actually move?

We can now consider the actual size of the atomic displacement. First let us consider the size of the normal mode coordinate for a single phonon in the high-temperature limit, equation (4.19). For a temperature $T = 300$ K and a frequency $\omega/2\pi = 2$ THz, the root-mean-square value of Q is about 5×10^{-14} $\mathrm{kg}^{1/2}$ Å. For a crystal with a monatomic unit cell and N atoms in the crystal, the mean-square atomic displacement is given as

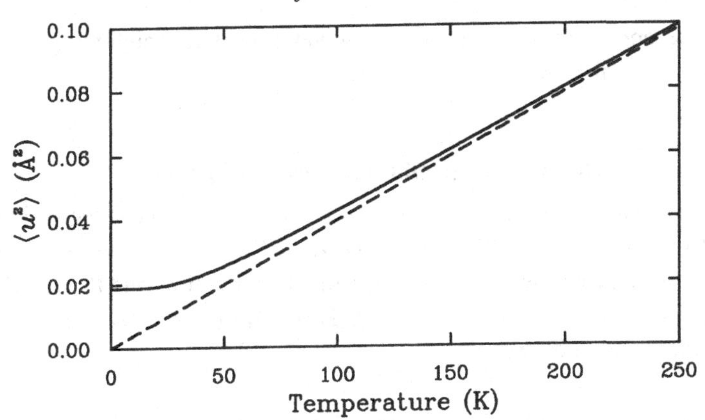

Figure 4.2: The full curve shows the calculated mean-squared displacement for an atom of mass 40 a.m.u. in a monatomic unit cell, with the approximation that all phonons have the same frequency of 2 THz. Note the small non-zero value at 0 K, and the linear temperature dependence at high temperatures. For comparison the broken curve shows the same quantity calculated using the classical high-temperature approximation.

$$\left\langle |\mathbf{u}(j)|^2 \right\rangle = \frac{1}{Nm_j} \sum_{\mathbf{k},\nu} \left\langle |Q(\mathbf{k},\nu)|^2 \right\rangle \tag{4.20}$$

This equation is derived from equation (4.2) in Appendix B. If we assume that all vibrations have the same angular frequency[1] ω, and note that there will be $3N$ modes, the root-mean-square value for the atomic displacement in the high-temperature limit can be written as

$$\left\langle |\mathbf{u}(j)|^2 \right\rangle = \frac{3k_B T}{m_j \omega^2} \tag{4.21}$$

For a frequency of 2 THz and an atomic mass of 40 a.m.u., the root-mean-square displacement at 300 K is calculated to be ~0.34 Å. This value is of typical order of magnitude measured in crystal structure refinements. Two points should be noted: firstly, heavier atoms vibrate with smaller amplitudes than light atoms; secondly, the mean-square displacement varies linearly with T at high temperatures (in the classical limit), as is generally observed experimentally. The temperature dependence of the mean-squared displacement is shown in Figure 4.2 for a range of temperatures, calculated without making the high-

[1] The approximation that all phonons have the same frequency is known as the *Einstein model*, and is discussed in more detail in Chapter 5. Whilst it is a crude approximation, and is unrealistic for acoustic modes, with a suitable choice of the average frequency the Einstein model can give a surprisingly good representation of real behaviour.

temperature approximation. The range of validity of the high-temperature approximation is apparent from this figure.

The crystallographic temperature factor

As we noted in the first chapter, the motions of atoms can be detected in an X-ray diffraction experiment through the *Debye–Waller* or *temperature factor*, which occurs in the expression for the structure factor (Willis and Pryor 1975; Castellano and Main 1985; Kuhs 1992). Formally the structure factor for Bragg scattering can be written as:

$$F(\mathbf{Q}) = \left\langle \sum_j f_j(Q)\exp\left(i\mathbf{Q}\cdot\mathbf{r}_j\right)\right\rangle$$

$$= \sum_j f_j(Q)\exp\left(i\mathbf{Q}\cdot\langle\mathbf{r}_j\rangle\right)\exp\left(-\frac{1}{2}\left\langle\left[\mathbf{Q}\cdot\mathbf{u}_j\right]^2\right\rangle\right) \tag{4.22}$$

where

$$\mathbf{Q} = h\mathbf{a}^* + k\mathbf{b}^* + l\mathbf{c}^* \tag{4.23}$$

and $f_j(Q)$ is the amplitude for scattering from an individual atom (known as the *atomic scattering factor* for X-ray diffraction and *scattering length* for neutron scattering). The anisotropic temperature factor (Willis and Pryor 1975) for a crystal with orthogonal unit cell axes is given as

$$\left\langle\left[\mathbf{Q}\cdot\mathbf{u}_j\right]^2\right\rangle = h^2a^{*2}\left\langle u_1^2\right\rangle + k^2b^{*2}\left\langle u_2^2\right\rangle + l^2c^{*2}\left\langle u_3^2\right\rangle$$

$$+2hka^*b^*\left\langle u_1u_2\right\rangle + 2hla^*c^*\left\langle u_1u_3\right\rangle + 2klb^*c^*\left\langle u_2u_3\right\rangle \tag{4.24}$$

where $\mathbf{u}_j = (u_1, u_2, u_3)$.

From the theory outlined above, we can write down an expression for the anisotropic temperature factor in terms of the normal mode coordinates:

$$\left\langle u_\alpha(j)u_\beta(j)\right\rangle = \frac{1}{Nm_j}\sum_{\mathbf{k},\nu}\langle Q(\mathbf{k},\nu)Q(-\mathbf{k},\nu)\rangle e_\alpha(j,\mathbf{k},\nu)e_\beta(j,-\mathbf{k},\nu) \tag{4.25}$$

It is therefore expected, as observed, that the temperature factor will increase linearly with temperature except at low temperatures.

Summary

1 We have introduced the formalism for the mode amplitudes in terms of the normal mode coordinates, and the harmonic energy of the crystal has been written in terms of these new variables.
2 The effects of quantum mechanics have been introduced into the picture. The lattice vibration quantum has been called the phonon. An expression for the average number of phonons at any temperature has been used in our energy equations.
3 The amplitudes of the vibrational modes have been related to the energy of the mode.
4 The energy of the mode has been shown to be related to the frequency of the mode and temperature.
5 We have derived an expression for the crystallographic temperature factor.

FURTHER READING

Ashcroft and Mermin (1976) ch. 23
Born and Huang (1954) ch. 4,15–16,38
Brüesch (1982) pp 27–33, 69–72
Willis and Pryor (1975) ch. 4–6

5

Lattice dynamics and thermodynamics

The results of the previous chapter are used to construct the lattice dynamics contributions to the major thermodynamic functions. The density of states method for the practical evaluation of these functions is described. The Einstein and Debye models for the density of states are discussed, and they are applied to the calculation of the heat capacity. The main results are also applied to the study of reconstructive phase transitions.

The basic thermodynamic functions

In Chapter 4 we obtained the following expression for the harmonic phonon energy of the crystal:

$$E = \sum_{\mathbf{k},\nu} \hbar\omega(\mathbf{k},\nu)\left[\frac{1}{2} + n(\omega,T)\right]$$

$$= \sum_{\mathbf{k},\nu} \hbar\omega(\mathbf{k},\nu)\left[\frac{1}{2} + \left(\exp\left(\hbar\omega(\mathbf{k},\nu)/k_{\mathrm{B}}T\right) - 1\right)^{-1}\right] \tag{5.1}$$

This is equivalent to the *internal energy*; the harmonic model does not allow for thermal expansion, so therefore this function is for constant volume. The constant volume heat capacity, C_V, is then equal to

$$C_V = \left(\frac{\partial E}{\partial T}\right)_V$$

$$= \sum_{\mathbf{k},\nu} \hbar\omega(\mathbf{k},\nu)\frac{\partial n(\omega,T)}{\partial T}$$

$$= \sum_{\mathbf{k},\nu} k_{\mathrm{B}}\left(\frac{\hbar\omega(\mathbf{k},\nu)}{k_{\mathrm{B}}T}\right)^2 \frac{\exp(\hbar\omega/k_{\mathrm{B}}T)}{\left[\exp(\hbar\omega/k_{\mathrm{B}}T) - 1\right]^2} \tag{5.2}$$

In the high-temperature limit, i.e. $k_BT \gg \hbar\omega$, E and C_V simply become equal to $3NZk_BT$ and $3NZk_B$ respectively, where Z is the number of atoms in the unit cell, and N is the number of unit cells in the crystal. This latter result is the classical *Dulong–Petit* result, which was found empirically in the early 19th century (Dulong and Petit 1819) and which is consistent with classical statistical mechanics. However, as T falls to zero, the heat capacity is observed also to fall to zero. In fact, before the effects of quantum mechanics were appreciated, the Dulong–Petit value for the heat capacity was the only value that could be calculated with the knowledge of the day. The failure of this result for low temperatures could not be understood until classical mechanics was superseded by quantum mechanics – the results of this chapter represent one of the most significant advances in the understanding of the crystalline state.

Using statistical thermodynamics any thermodynamic quantity can be obtained from the partition function, Z, which following the definition in Appendix C is given as

$$Z = \exp(-\varphi / k_BT)\prod_{\mathbf{k},\nu} \frac{\exp(-\hbar\omega(\mathbf{k},\nu)/2k_BT)}{1-\exp(-\hbar\omega(\mathbf{k},\nu)/k_BT)} \tag{5.3}$$

where φ is the potential energy of the crystal. The free energy at constant volume, F (the *Helmholtz* free energy), is obtained from:

$$\begin{aligned} F &= -k_BT\ln Z \\ &= \varphi + \frac{1}{2}\sum_{\mathbf{k},\nu}\hbar\omega(\mathbf{k},\nu) + k_BT\sum_{\mathbf{k},\nu}\ln\left[1-\exp(-\hbar\omega(\mathbf{k},\nu)/k_BT)\right] \\ &= \varphi + k_BT\sum_{\mathbf{k},\nu}\ln\left[2\sinh(\hbar\omega(\mathbf{k},\nu)/2k_BT)\right] \end{aligned} \tag{5.4}$$

The two forms for the free energy are given since they are both frequently encountered in the literature. The entropy S follows from the first of the two equations for F as

$$\begin{aligned} S = -\frac{\partial F}{\partial T} &= -k_B\sum_{\mathbf{k},\nu}\ln\left[1-\exp(-\hbar\omega(\mathbf{k},\nu)/k_BT)\right] - \frac{1}{T}\sum_{\mathbf{k},\nu}\hbar\omega(\mathbf{k},\nu)n(\omega,T) \\ &= \frac{1}{2T}\sum_{\mathbf{k},\nu}\hbar\omega_\nu(\mathbf{k})\coth(\hbar\omega_\nu(\mathbf{k})/2k_BT) - k_B\sum_{\mathbf{k},\nu}\ln\left[2\sinh(\hbar\omega(\mathbf{k},\nu)/2k_BT)\right] \end{aligned} \tag{5.5}$$

These results have three main applications. Firstly, they enable us to understand the observed thermodynamic properties of solids, and we will develop models that can explain the quantitative details of experimental data (e.g. the behaviour of the heat capacity at low temperatures). Secondly, the results can be used as a predictive tool for calculating the thermodynamic properties of materials within environments that cannot be reproduced in the laboratory. Thirdly, we can can use these results to understand crystal stability and phase transitions.

Evaluation of the thermodynamic functions and the density of states

General considerations

The equations for the thermodynamic functions can in principle be evaluated directly using a long list of frequency values taken from the dispersion curves calculated over a fine grid of wave vectors within the first Brillouin zone. Since the energy of a vibration depends only on its frequency there is some advantage in developing a formalism that relies only on the frequency distribution. We define a quantity called the *density of states*, $g(\omega)$, such that the number of modes with angular frequencies between ω and $\omega + \mathrm{d}\omega$ is equal to $g(\omega)\mathrm{d}\omega$. Thus the harmonic phonon energy of the crystal can be written in the new form:

$$E = \int \hbar\omega g(\omega)\left[\frac{1}{2} + n(\omega, T)\right]\mathrm{d}\omega \tag{5.6}$$

The heat capacity, for example, can then be written as:

$$C_V = \frac{\partial}{\partial T}\int \hbar\omega g(\omega) n(\omega, T)\mathrm{d}\omega \tag{5.7}$$

In many cases, however, a detailed model for the calculation of the dispersion curves is not available. This is often a problem when trying to construct the thermodynamic functions of minerals that are unstable at room temperature and pressure. As a result, there are a number of schemes for approximating the form of $g(\omega)$ for the interpretation of experimental data or the prediction of thermodynamic quantities.

Optic modes

The simplest approach to modelling the contribution of the optic branches to the density of states is to make the assumption that all the optic modes have the same frequency, whose value can be determined by fitting the calculated thermodynamic functions against experimental data. This approach is known as the Einstein model (Einstein 1907). Although this seems like a rather drastic simplification, it turns out that with a suitable choice of the mean frequency value – called the *Einstein frequency*, ω_{E} – the model can reproduce the heat capacity over a wide range of temperatures. This is often extended to include the acoustic modes with the same mean frequency value.

The next level of sophistication is to take a small number of representative frequency values. This may be useful for systems that are so complex that the accurate calculation of the density of states is computationally impractical, or for cases when a good model is not available but when experimental spectroscopic data have been obtained (Salje and Werneke 1982). If one took the values for $\mathbf{k} = 0$ only, as given by spectroscopic measurements, the calculations would neglect any effects due to dispersion. Baldereschi (1973) has shown that there is a special point within any Brillouin zone for which the calculated frequencies give accurate mean values for thermodynamic analysis. With this approach calculations only need be performed at a single wave vector.

Finally, the density of states for the optic modes can be modelled assuming a uniform distribution of frequencies between two cut-off values for the lower-frequency modes, and delta functions for the high-frequency bond-stretching vibrations. The acoustic modes can be incorporated using the Debye model described below. The resultant density of states is often a good approximation for relatively complex crystals. The cut-off values can be obtained from spectroscopic measurements of the phonon frequencies at $\mathbf{k} = 0$. This method is often used for materials in non-ambient environments (e.g. high pressure), when calorimetric experiments are impossible but when spectroscopy can be used. In some cases the model can be refined by using two or more such distributions when experimental data indicate significant separations between bands of modes. This approach was largely pioneered by Kieffer (1979a–c, 1980), and has been successfully applied to a wide range of minerals – it is, of course, known as the *Kieffer model*.

A number of different models to represent the density of states in the complex aluminosilicate andalusite, Al_2SiO_5, are compared in Figure 5.1.

Acoustic modes

To calculate the contribution to the density of states from the acoustic modes we will assume that the frequency dispersion is linear with wave vector, and for simplicity we will also assume that we are able to replace the slopes of each branch with the average slope.[1] This approach was introduced by Debye (1912), and is known as the *Debye model*.[2] We define an average velocity of sound, c, such that

[1] In practice this assumption is not necessary, but for our purposes it keeps the detail simple.

[2] It is interesting to note that the Einstein and Debye models were proposed before the atomic structure of crystals had been experimentally verified (Bragg 1913) and before the early work on lattice dynamics by Born and von Kármán (1912, 1913).

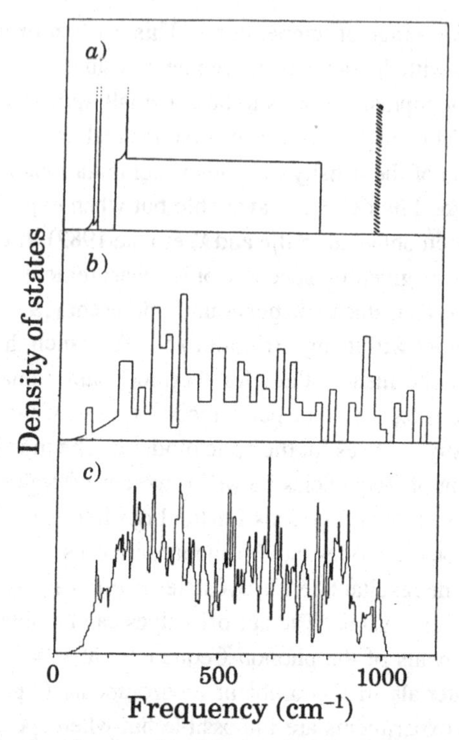

Figure 5.1: The density of states of andalusite, Al_2SiO_5, calculated using three models: *a)* the Kieffer model (Kieffer 1980); *b)* using spectroscopic optic mode frequencies and a Debye model for the acoustic modes (Salje and Werneke 1982); *c)* a lattice dynamics calculation using a fine grid of wave vectors in reciprocal space (Winkler and Buehrer 1990).

$$\omega = ck \tag{5.8}$$

The volume of the Brillouin zone is equal to $8\pi^3 N/V$, where V is the volume of the crystal. Remembering that there are N points in the Brillouin zone, the density of points in reciprocal space will be equal to $V/8\pi^3$. The number of points with wave vector between k and $k + \mathrm{d}k$ will then be equal to

$$g(k)\mathrm{d}k = \frac{V}{8\pi^3} 4\pi k^2 \mathrm{d}k \tag{5.9}$$

We can convert this expression from wave vector to frequency ($\omega = ck$, $\mathrm{d}\omega = c\mathrm{d}k$) to obtain an expression for the density of states:

$$g(\omega)\mathrm{d}\omega = \frac{3V}{8\pi^3} 4\pi \left(\frac{\omega}{c}\right)^2 \frac{\mathrm{d}\omega}{c} \tag{5.10}$$

where the extra factor of 3 comes from the fact that there are three acoustic modes for each wave vector. Note that this gives the general result that the density of states is proportional to ω^2 in the limit that $\omega \propto k$ as $\omega \to 0$. Thus the contribution of the acoustic phonons to the thermal energy, equation (5.6), is

$$E = \int_0^{\omega_D} \left(\frac{3V\hbar\omega^3}{2\pi^2 c^3} \right) \left[\exp(\hbar\omega / k_B T) - 1 \right]^{-1} d\omega \tag{5.11}$$

where we have now dropped the constant zero-point energy from our equations. We have defined a cut-off frequency, ω_D, called the *Debye frequency*. The Debye frequency can be calculated from an effective cut-off wave vector determined by the number of points within a sphere in reciprocal space:

$$\frac{Vk^3}{6\pi^2} = N \Rightarrow \omega_D = c\left(\frac{6\pi^2 N}{V} \right)^{1/3} \tag{5.12}$$

Alternatively, the value of ω_D can be treated as an adjustable parameter to be determined by fitting to experimental data. In this case we re-write equation (5.11) as

$$E = 3N\int_0^{\omega_D} 3\hbar \left(\frac{\omega}{\omega_D} \right)^3 \left[\exp(\hbar\omega / k_B T) - 1 \right]^{-1} d\omega \tag{5.13}$$

ω_D can be treated as a constant, or can be assumed to be weakly temperature dependent.

If we now make the substitutions

$$x = \hbar\omega / k_B T \; ; \; x_D = \hbar\omega_D / k_B T \tag{5.14}$$

we obtain the following expression for the thermal energy:

$$E = \left(\frac{3V\hbar}{2\pi^2 c^3} \right) \left(\frac{k_B T}{\hbar} \right)^4 \int_0^{x_D} x^3 \left(e^x - 1 \right)^{-1} dx \tag{5.15}$$

This integral must be solved numerically. However, in the low-temperature limit we can allow the upper limit on the integral, x_D, to go to infinity, since it will already have a very large value. This gives a standard integral with the solution:

$$\int_0^{\infty} x^3 \left(e^x - 1 \right)^{-1} dx = \frac{\pi^4}{15} \tag{5.16}$$

Thus the thermal energy due to the acoustic modes at low temperature is equal to

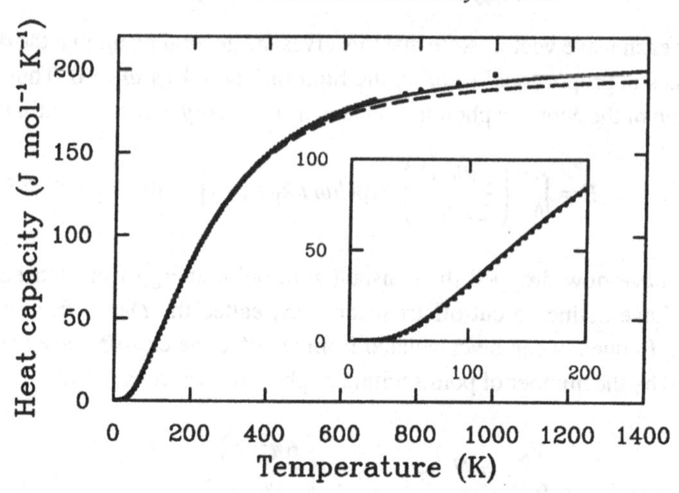

Figure 5.2: Heat capacity of andalusite (Al_2SiO_5) showing the experimental data for C_P (full circles, Robie and Hemingway 1984; Hemingway et al. 1991), the results from a lattice dynamics calculation of C_V (dashed curve, Winkler et al. 1991a; Winkler and Buehrer 1990), and the calculated form of C_P obtained from the calculation of C_V using equation (5.24) (continuous curve).

$$E = \left(\frac{3V\hbar}{2\pi^2 c^3}\right)\left(\frac{k_B T}{\hbar}\right)^4\left(\frac{\pi^4}{15}\right) = \frac{V\pi^2 (k_B T)^4}{10(c\hbar)^3} \tag{5.17}$$

The heat capacity

The heat capacity is the easiest thermodynamic quantity to measure experimentally, and is worthy of a more detailed consideration. An example of the heat capacity of a complex system, andalusite, is given in Figure 5.2, in which the experimental values of C_P (rather than C_V) are compared with calculations using the density of states obtained from a lattice dynamics calculation as shown in Figure 5.1c. We can explain the characteristic form of the temperature dependence of the heat capacity in three stages.

Classical model

We have already quoted above the classical Dulong–Petit result for the heat capacity, namely

$$C_V = 3NZk_B \tag{5.18}$$

This result is obeyed at high temperatures for all systems (neglecting the contribution from the electrons), although anharmonic effects may tend to lower the value of the heat capacity slightly. For many systems the Dulong–Petit regime may be rather higher than room temperature, particularly if there are high-frequency vibrations, so that a full quantum-mechanical calculation of the heat capacity is then required. For andalusite the value of C_V obtained from equation (5.18) is 199.5 J mol^{-1} K^{-1}, and from Figure 5.2 it can be seen that the Dulong–Petit regime for andalusite is above 1000 K.

Einstein model

On cooling, the heat capacity is found to decrease in value from the Dulong–Petit limit, eventually reaching zero at 0 K. As we have described above, Einstein (1907) suggested making the simple approximation that all the frequencies be set equal to just one value, ω_E. The heat capacity will then be equal to

$$C_V = 3NZk_B\left(\frac{\hbar\omega_E}{k_BT}\right)^2 \frac{\exp(\hbar\omega_E / k_BT)}{\left[\exp(\hbar\omega_E / k_BT)-1\right]^2} \tag{5.19}$$

This is known as the *Einstein model* for the heat capacity. It may be crude, but it gives a reasonable picture; in particular, the Einstein model predicts the decrease of the heat capacity on cooling. The main purpose of the Einstein model was to demonstrate the fact that the temperature dependence of the heat capacity arises from quantum-mechanical effects, rather than to give quantitative results. When equation (5.19) is fitted against experimental data, the model generally agrees with the data quite well, giving a useful estimate for the appropriate value of ω_E. However, the model does not correctly reproduce the behaviour very close to absolute zero: the calculated heat capacity falls to zero on decreasing temperature faster than is observed experimentally. The reason for this is that at low temperature the acoustic modes with small wave vectors, and hence low frequencies, will be the most populated modes, and these are essentially neglected in the Einstein model. So we need to turn to a different model for the acoustic mode contribution to the heat capacity.

Debye model

The Debye (1912) model for the heat capacity uses the acoustic mode density of states (equation (5.10)). The heat capacity for this model can be obtained from equations (5.2) and (5.13) for the internal energy, giving the general formula:

$$C_V = 3N\int_0^{\omega_D} 3\hbar\left(\frac{\omega}{\omega_D}\right)^3 \frac{\partial n(\omega,T)}{\partial T}\,d\omega \tag{5.20}$$

At low temperature, we can use equation (5.17) to obtain the simple result:

$$C_V = \left(\frac{2k_B V\pi^2}{5}\right)\left(\frac{k_B T}{c\hbar}\right)^3 \tag{5.21}$$

The low-temperature heat capacity is thus predicted to vary as the third power of the temperature, and this behaviour is observed experimentally in many systems over a range of a few degrees above absolute zero. It is common practice to make use of a particular temperature called the *Debye temperature*, Θ_D, defined as

$$\frac{\hbar\omega_D}{k_B T} = \frac{\Theta_D}{T} \Rightarrow \Theta_D = \frac{c\hbar}{k_B}\left(\frac{6\pi^2 N}{V}\right)^{1/3} \tag{5.22}$$

The low-temperature heat capacity can be re-expressed in terms of the Debye temperature:

$$C_V = \frac{12\pi^4 N k_B}{5}\left(\frac{T}{\Theta_D}\right)^3 \approx 234 N k_B \left(\frac{T}{\Theta_D}\right)^3 \tag{5.23}$$

Because the Debye model neglects any curvature of the acoustic mode dispersion curves, we expect that the Debye temperature will have a weak dependence on the temperature at which it is evaluated. In general application, a single Debye frequency can be determined by fitting the calculated heat capacity, equation (5.20), against experimental data over a range of temperatures, to give a reasonable but not perfect agreement at any temperature, or else the Debye frequency can be fitted at every temperature to give perfect agreement at all temperatures but with a temperature-dependent Debye frequency.

Conversion to constant pressure

The theory outlined above for harmonic crystals is applicable only under the condition of constant volume: the harmonic model does not predict the existence of thermal expansion. However, all experiments are performed under the condition of constant pressure, since it is virtually impossible to design an experiment that will prevent thermal expansion. There is a simple conversion between the heat capacity at constant pressure, C_P, and C_V (e.g. Adkins 1975, p 115):

$$C_P = C_V + TV\beta^2 / K_T \tag{5.24}$$

where β is the thermal expansion coefficient,

$$\beta = \frac{1}{V}\left(\frac{\partial V}{\partial T}\right)_P = K_T\left(\frac{\partial P}{\partial T}\right)_V \tag{5.25}$$

and K_T is the isothermal compressibility,

$$K_T = -\frac{1}{V}\left(\frac{\partial V}{\partial P}\right)_T \tag{5.26}$$

The compressibility can be calculated from the elastic constants (Nye 1964), which can themselves be calculated by lattice energy programs if experimental values are not available:

$$K_T = \sum_{i,j=1,3} S_{ij} \tag{5.27}$$

where **S** is the elastic compliance matrix, and is given by the inverse of the elastic constant tensor, $\mathbf{C}^{-1}$ (Nye 1964). On the other hand, the thermal expansion coefficient is not so straightforward to obtain by calculation. Thermal expansion is an anharmonic effect, but for the present purposes can be treated within the spirit of the harmonic approximation. We first note that the pressure, P, is given by the derivative of the Helmholtz free energy with respect to the volume V. From equation (5.4) we have

$$P = -\left(\frac{\partial F}{\partial V}\right)_T = -\frac{\partial \varphi}{\partial V} - \frac{1}{2}\sum_{\mathbf{k},\nu}\hbar\frac{\partial \omega(\mathbf{k},\nu)}{\partial V} - \sum_{\mathbf{k},\nu} n(\omega,T)\hbar\frac{\partial \omega(\mathbf{k},\nu)}{\partial V} \tag{5.28}$$

Inserting this expression into the equation for the thermal expansion coefficient yields:[3]

$$\beta = -K_T\sum_{\mathbf{k},\nu}\hbar\frac{\partial \omega(\mathbf{k},\nu)}{\partial V}\frac{\partial n(\omega,T)}{\partial T} \tag{5.29}$$

This can be represented in a manner which reflects the similarity to the heat capacity. We define a quantity that gives the contribution of each mode to the heat capacity:

$$\tilde{C}_{\mathbf{k},\nu} = \hbar\omega(\mathbf{k},\nu)\frac{\partial n(\omega,T)}{\partial T} \tag{5.30}$$

We also define a quantity known as the *mode Grüneisen parameter*:

[3] Note that we assume that the frequencies are independent of temperature for constant volume, and depend on temperature only through the effects of thermal expansion. We will consider the intrinsic temperature dependence of the frequencies in Chapter 8.

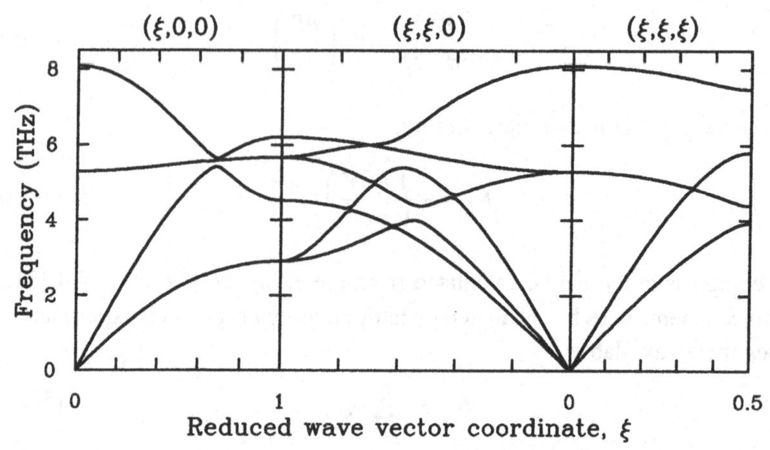

Figure 5.3: Calculated phonon dispersion curves for NaCl, using the model of Sangster and Atwood (1978).

$$\gamma_{\mathbf{k},\nu} = -\frac{V}{\omega(\mathbf{k},\nu)}\frac{\partial\omega(\mathbf{k},\nu)}{\partial V} = -\frac{\partial(\ln\omega(\mathbf{k},\nu))}{\partial(\ln V)} \tag{5.31}$$

and the related *mean Grüneisen parameter*:

$$\gamma = \sum_{\mathbf{k},\nu}\gamma_{\mathbf{k},\nu}\tilde{C}_{\mathbf{k},\nu} / C_V \tag{5.32}$$

where the negative sign in (5.31) gives a positive contribution to γ: it is generally found that frequencies tend to decrease in value as the volume increases, which one might expect since interatomic forces will decrease as bond lengths increase.

The mean Grüneisen parameter may be expected to be temperature dependent (Barron et al. 1980), although for some systems (notably for monatomic unit cells) γ does not vary significantly with temperature. For the Debye model, every mode frequency is simply proportional to the Debye frequency, so that each mode Grüneisen parameter has the same value, and the heat capacity terms in equation (5.32) cancel.

Finally the thermal expansion coefficient can then be written as

$$\beta = \frac{K_T\gamma C_V}{V} \tag{5.33}$$

Each of the quantities in this expression are calculable, although the evaluation of γ requires several rather lengthy calculations over a grid in reciprocal space performed using different values of the unit cell volumes. Thus from a harmonic lattice dynamics calculation it is possible to estimate the thermal

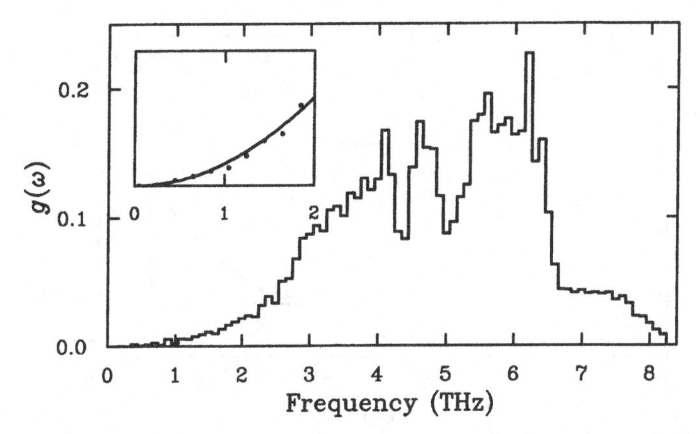

Figure 5.4: Calculated density of states for NaCl. The inset shows the behaviour at small ω, where the curve is a parabola fitted to the density of states. The scatter on the histogram reflects the use of a finite grid size in the evaluation of the density of states.

expansion and hence the conversion from C_V to C_P. It may, however, often be preferable to use experimental data for the conversion factor!

Worked example: the heat capacity of NaCl

NaCl is a good simple example to illustrate the methods described above. The dispersion curves calculated with a model developed by Sangster and Atwood (1978) are shown in Figure 5.3; they can be compared with the experimental results shown in Figure 3.6a. The density of states for this model, evaluated with a square grid in reciprocal space of $0.1a^*$ between grid points, is shown in Figure 5.4. Note the ω^2 dependence at small ω, which is highlighted in the inset to Figure 5.4. The heat capacity calculated from this density of states is given in Figure 5.5. The T^3 dependence at low temperatures is highlighted in the inset to Figure 5.5.

In Figure 5.6 we compare the heat capacity with a calculation using the Einstein model with a value of ω_E chosen to give reasonable agreement at higher temperatures. The detail at low temperature is highlighted in the inset to Figure 5.6. We see that the Einstein model provides a reasonable description of the heat capacity for temperatures above 50 K, and in particular it correctly reproduces the departures from the Dulong–Petit limit. The value of ω_E can only be obtained by fitting. In some senses ω_E gives an average of all the frequencies appropriately weighted. For the calculations of some other thermodynamic quantity using the Einstein model the best value of ω_E may well be different, reflecting a different weighting.

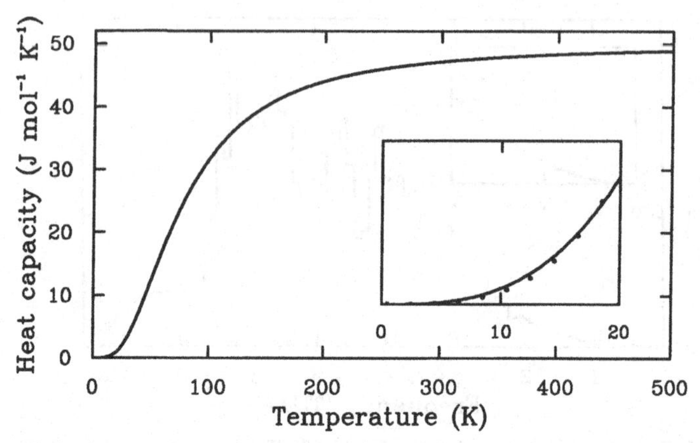

Figure 5.5: The calculated heat capacity of NaCl. The inset shows the behaviour at low temperature, where the curve is a fitted T^3 function and the filled circles are the actual calculated data.

Technically the Debye model does not apply to crystals such as NaCl, because half the modes are optic modes. However, some authors do try to fit the Debye model to measured heat capacities. In one sense this may be possible for NaCl, if one considers the optic modes to be extensions of the acoustic modes, but this justification is not valid for more complex crystals. Accordingly we do not pursue this analysis here. In any case, the correction for the acoustic modes can be incorporated into a Kieffer model.

Free energy minimisation methods and the quasi-harmonic approximation

The use of the Grüneisen parameters suggests that it may be possible to incorporate anharmonic interactions into calculations of structure and lattice dynamics via the dependence of the frequencies on volume and structure. The equilibrium structure at any temperature is always that with the lowest free energy. This suggests that it may be possible to model temperature dependence of a crystal structure by calculating the minimum of the free energy, equation (5.4). The energy advantage of having as small a volume as possible is offset at higher temperatures by the entropy advantage of lower frequencies, leading to a net thermal expansion. The assumption that the anharmonicity is restricted to thermal expansion, so that the temperature dependence of the phonon frequencies arises only from the dependence on crystal structure and volume, is called the *quasi-harmonic approximation*, since the lattice dynamics are still treated within the harmonic approximation. Whilst this model is good for predicting

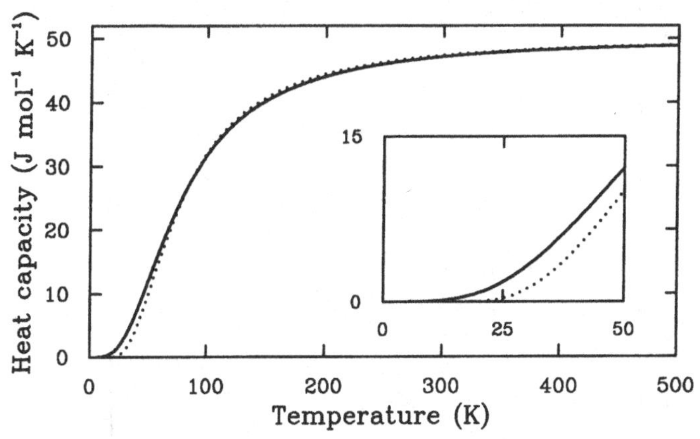

Figure 5.6: Comparison of the Einstein model for the heat capacity of NaCl with the exact result. The inset shows the comparison in more detail at low temperature.

thermal expansion, it does neglect any intrinsic temperature dependence of the phonon frequencies. For the calculation of thermal expansion this may not be a problem, but the quasi-harmonic model described here is unable to model displacive phase transitions which arise primarily through the intrinsic anharmonic interactions (Chapter 8). The free energy minimisation approach using the quasi-harmonic approximation has been applied to the study of high-temperature–high-pressure phenomena in minerals (Parker and Price 1989).

Reconstructive phase transitions

The stability of any crystal phase against transformation to another phase is determined by the difference between the free energies of the two phases, which is itself temperature dependent. The phase transition temperature at a constant pressure can be calculated by comparing free energy curves constructed using equation (5.4); the temperature at which the curves cross will be the transition temperature. To calculate the complete pressure–temperature phase diagram the Clausius–Clapeyron equation can be used, which relates the slope of the phase diagram, $\mathrm{d}P/\mathrm{d}T$, to the differences in the entropy (ΔS) and volume (ΔV) across the phase boundary (Adkins 1975, pp 188–189; Mandl 1971, pp 231–234):

$$\frac{\mathrm{d}P}{\mathrm{d}T} = \frac{\Delta S}{\Delta V} \tag{5.34}$$

The change in volume can easily be obtained from lattice energy calculations. The change in entropy can be calculated from a lattice dynamics calculation

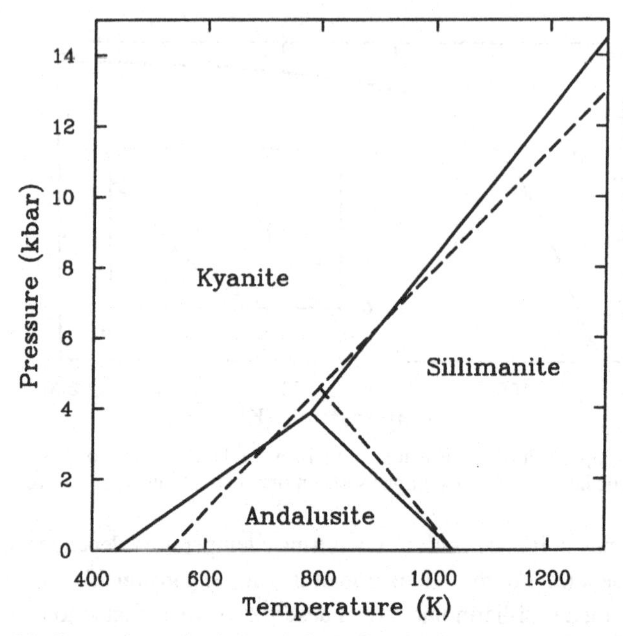

Figure 5.7: Comparison of calculated phase diagram for Al_2SiO_5 (dashed lines) and the experimentally determined phase boundary lines (continuous lines) taken from Hemingway et al. (1991). It should be noted that the phase boundaries are particularly sensitive to Al–Si disorder and the degree of crystallinity.

via equation (5.5). It can be demonstrated from a consideration of equation (5.5) that at high temperatures ΔS is virtually independent of temperature. Similarly ΔV will also be relatively independent of temperature if the thermal expansion coefficients of both phases are similar. Therefore equation (5.34) predicts that the phase boundary will be approximately straight, and a single calculation of ΔS and ΔV will yield the full phase diagram

Examples of the calculations of phase diagrams are for Mg_2SiO_4 (Price et al. 1987b) and Al_2SiO_5 (Winkler et al. 1991a). The work on the polymorphs of Mg_2SiO_4 is a good example of what can be achieved with a reliable model interatomic potential. The calculated structure and vibrational frequencies for forsterite are in good agreement with experimental data. From a calculated density of states the heat capacity, Grüneisen parameter, and thermal expansion coefficient were calculated. Finally, the structures and free energies of all three polymorphs were calculated, from which the P–T phase diagram was constructed. The final phase diagram is not in exact agreement with experimental data, but it is clear how the model could be improved to remove this discrepancy.

The phase diagram for Al_2SiO_5 is reproduced in Figure 5.7. This was calculated using an extrapolation procedure for the evaluation of the phase boundaries. The limits of the model required that two fixed state points needed to be taken from experimental data. The major success of this model was the calculation of the triple point observed in this system, which was found to be in qualitative agreement with experimental data (Bohlen et al. 1991; Hemingway et al. 1991). From the examples that have appeared in the literature it appears that the slopes of phase boundaries can be calculated with higher accuracy than the positions of fixed points.

Summary

1 We have obtained expressions for the major thermodynamic functions within the harmonic approximation.
2 The concept of the phonon density of states has been introduced, and methods for its calculation have been described.
3 A number of different approaches to the calculation of the heat capacity have been described. Different approaches are relevant for different temperature regimes.
4 Thermal expansion coefficients can be estimated from lattice dynamics calculations. This is important for converting calculations performed under constant volume to constant pressure results for comparison with experimental data.
5 The models give reasonable results for the phase boundaries associated with reconstructive phase transitions.

FURTHER READING

Ashcroft and Mermin (1976) ch. 23
Born and Huang (1954) ch. 4,6,41,43
Brüesch (1982) pp 45–51
Cochran (1973) ch. 6
Kittel (1976) ch. 5
Mandl (1971) ch. 6

6
Formal description

The standard formal methods of lattice dynamics are developed. As an introduction to the formal methods we return to the diatomic chain and recast the equations of motion in a more general form. The dynamical matrix is introduced, and it is shown that the frequencies are obtained from the eigenvalues of the dynamical matrix and that the atomic motions are given by the eigenvectors. The Hamiltonian is written in terms of the normal mode coordinates. The casual reader does not need this chapter for the rest of the book, and may be advised to skip over the details.

Review and problems

So far we have worked with rather simple models, which have given us a number of results that are of general use. We have found, however, that the calculations get exceedingly difficult even when we get as complicated as the diatomic unit cell. The problems with our simple approach are the following:

- The simple approach only works in high-symmetry cases, where any mode will contain only one direction of motion. In general there will be mixing of different motions in each vibrational mode, and different modes may correspond to different mixtures of the same atomic motions.
- The equations of motion are difficult to solve when there are many (i.e. more than two) atoms in the unit cell.
- The equations get cumbersome when we include forces from distant neighbours.
- We have not explicitly considered how to treat atoms in general positions in the unit cell.

For these reasons we need a general theoretical framework to define the basic lattice dynamics problem. This can then be used in a computer program to calculate the dispersion curves and the mode eigenvectors, which form a matrix of the linear combinations of cartesian atomic displacements.

The diatomic chain revisited

The basic ideas that we will develop in this chapter can be illustrated by considering the one-dimensional diatomic model discussed in Chapter 3. We will progress by re-writing the basic equations of motion. We will first make a change of variables:

$$E = M^{1/2}\tilde{U}_k \ ; \ e = m^{1/2}\tilde{u}_k \tag{6.1}$$

so that we will solve our equations for (E, e) instead of (U, u). The equations of motion in matrix form are now:

$$\begin{pmatrix} E \\ e \end{pmatrix}\omega_k^2 = \mathbf{D}(\mathbf{k})\begin{pmatrix} E \\ e \end{pmatrix} \tag{6.2}$$

where the matrix $\mathbf{D}(\mathbf{k})$, which is known as the *dynamical matrix,* is given by:

$$\mathbf{D}(\mathbf{k}) = \begin{pmatrix} \dfrac{(G+g)}{M} & -\dfrac{(G+g\exp(-ika))}{(Mm)^{1/2}} \\ -\dfrac{(G+g\exp(ika))}{(Mm)^{1/2}} & \dfrac{(G+g)}{M} \end{pmatrix} \tag{6.3}$$

We will see later that the dynamical matrix is of central importance in the calculation of dispersion curves. It has two symmetry properties that are apparent in our case, namely that $\mathbf{D}(-\mathbf{k}) = \mathbf{D}^*(\mathbf{k})$, and that the matrix is *Hermitian*, i.e. $\mathbf{D}^T(\mathbf{k}) = \mathbf{D}^*(\mathbf{k})$. This latter property gives the condition that the eigenvalues of $\mathbf{D}(\mathbf{k})$ are real, that is that the squares of the frequencies of the lattice vibrations are necessarily real. Equation (6.3) can be compared with equation (3.7).

Equations (6.2) and (6.3) have two general solutions:

$$\begin{aligned} &\textit{solution } 1\text{:} \quad \omega_1^2 \ , \ (E_1, e_1) \\ &\textit{solution } 2\text{:} \quad \omega_2^2 \ , \ (E_2, e_2) \end{aligned} \tag{6.4}$$

If we include both solutions in the matrix equation of motion (6.2), we obtain

$$\mathbf{e}\cdot\Omega = \mathbf{D}(\mathbf{k})\cdot\mathbf{e} \tag{6.5}$$

where the *frequency* and *displacement* matrices are, respectively, equal to

$$\Omega = \begin{pmatrix} \omega_1^2 & 0 \\ 0 & \omega_2^2 \end{pmatrix} \; ; \; \mathbf{e} = \begin{pmatrix} E_1 & E_2 \\ e_1 & e_2 \end{pmatrix} \tag{6.6}$$

The frequency matrix Ω can be obtained from the dynamical matrix $\mathbf{D}$ by a simple procedure. Equation (6.5) can be simply rearranged to give:

$$\Omega = \mathbf{e}^{-1} \cdot \mathbf{D} \cdot \mathbf{e} \tag{6.7}$$

As the frequency matrix Ω is diagonal by definition, all we have done is to *diagonalise* the dynamical matrix $\mathbf{D}$. The elements of the diagonal matrix are simply the eigenvalues of the diagonalised matrix, and the elements of the diagonalising matrix (in this case, the displacement matrix $\mathbf{e}$) are simply the eigenvectors of the matrix being diagonalised (here the dynamical matrix, $\mathbf{D}$). Thus we refer to the squares of the frequencies as the *eigenvalues* of the dynamical matrix, and the displacements produced by the corresponding modes of vibration as the *mode eigenvectors*. The solution to equation (6.7) contains an arbitrary scale factor on the eigenvectors, which are therefore defined to be normalised such that:

$$E_1^2 + e_1^2 = E_2^2 + e_2^2 = 1 \tag{6.8}$$

The eigenvectors therefore give the relative atomic displacements rather than their absolute values – we considered the magnitudes of the absolute displacements in Chapter 4. We also note that the eigenvectors are orthogonal, which is mathematically described by the condition:

$$E_1 E_2 + e_1 e_2 = 0 \tag{6.9}$$

The physical meaning is that different modes are independent, and can therefore be added linearly without interacting.

It is clear therefore that all the information that determines both the frequencies and displacements associated with the sets of vibrations of a system is contained in the dynamical matrix, and the task of calculating the dispersion curves resolves itself as the task of setting up the dynamical matrix. We shall see below that the dynamical matrix is determined by the force constants between a set of reference atoms and all their neighbours, as is clear in this example. The Hermitian character of the dynamical matrix is the condition that the eigenvalues are real, although they may be negative. Negative eigenvalues imply unphysical imaginary frequencies. The existence of imaginary mode frequencies implies that the crystal is unstable with respect to the distortion described by the corresponding eigenvector. Whilst this may often imply that a model used in a calculation gives an unstable crystal structure, the concept is of central importance in the theory of phase transitions. In general, calculations of dispersion curves are performed by evaluating the dynamical matrix, and obtaining the eigenvalues and eigenvectors. The general case can be rather complicated!

The equations of motion and the dynamical matrix

This part of the chapter gets rather complicated – actually, the concepts are not too difficult, but there are so many labels on everything that it all looks rather difficult. It is worth beginning by pointing out where we are going. In order to stay quite general, all the equations are in matrix form. You can expand things out to check that we aren't saying anything new – the expansions are not difficult to do. We will write the harmonic energy in matrix form, and then write the equation of motion in the same matrix form. We will take the same set of solutions as before, and insert them into the matrix equation of motion in order to get solutions for the frequencies. By putting everything together again, we will end up with the same type of matrix equation as in equation (6.7). In our nomenclature we will allow for any number of atoms in the unit cell, and will allow for interactions with many neighbouring atoms in other unit cells. We will work in three dimensions.

Let us start by defining the lattice energy, W, as a sum over all atom–atom interactions:

$$W = \frac{1}{2}\sum_{jj',ll'} \varphi\begin{pmatrix} jj' \\ ll' \end{pmatrix} \tag{6.10}$$

where j denotes an atom in the l-th unit cell, and the interaction energy φ is for the pair of atoms (jl) and $(j'l')$. The harmonic displacement energy is expressed in matrix form as:

$$E^{\text{harm}} = \frac{1}{2}\sum_{jj',ll'} \mathbf{u}^{\text{T}}(jl)\cdot\mathbf{\Phi}\cdot\mathbf{u}(j'l') = \frac{1}{2}\sum_{jj',ll'}\sum_{\alpha\beta} u_\alpha(jl)\Phi_{\alpha\beta}u_\beta(j'l') \tag{6.11}$$

where we define the 3×1 displacement matrix, $\mathbf{u}(jl)$, as:

$$\mathbf{u}(jl) = \begin{pmatrix} u_x(jl) \\ u_y(jl) \\ u_z(jl) \end{pmatrix} \tag{6.12}$$

and $\mathbf{u}^{\text{T}}$ is the transpose of $\mathbf{u}$.

The force constant matrix $\mathbf{\Phi}$ is a 3×3 matrix, with elements:

$$\Phi_{\alpha\beta}\begin{pmatrix} jj' \\ ll' \end{pmatrix} = \frac{\partial^2 W}{\partial u_\alpha(jl)\partial u_\beta(j'l')} \tag{6.13}$$

The subscripts α and β denote the cartesian vector components x, y and z. All that we have introduced is a compact way of writing the harmonic term in the Taylor expansion of the crystal energy. But in previous chapters we haven't quite written it this way. You can easily convince yourself, by writing out all the terms, that our previous notation is consistent with the following matrix equation:

$$E^{\text{harm}} = \frac{1}{4}\sum_{jj',ll'}\left[\mathbf{u}(jl) - \mathbf{u}(j'l')\right]^{\mathrm{T}} \cdot \boldsymbol{\phi}\begin{pmatrix} jj' \\ ll' \end{pmatrix} \cdot \left[\mathbf{u}(jl) - \mathbf{u}(j'l')\right] \tag{6.14}$$

where the matrix $\boldsymbol{\phi}$ has elements:

$$\phi_{\alpha\beta}\begin{pmatrix} jj' \\ ll' \end{pmatrix} = \frac{\partial^2 \varphi\begin{pmatrix} jj' \\ ll' \end{pmatrix}}{\partial u_\alpha(jl)\partial u_\beta(j'l')} \tag{6.15}$$

Equation (6.14) contains an extra factor $\frac{1}{2}$ (making a total factor of $\frac{1}{4}$) because of double counting backwards and forwards. As before, by expanding all the terms it is straightforward to show that equation (6.14) is consistent with equation (6.11), with

$$\Phi_{\alpha\beta}\begin{pmatrix} jj' \\ ll' \end{pmatrix} = -\phi_{\alpha\beta}\begin{pmatrix} jj' \\ ll' \end{pmatrix} + \delta_{jj'}\delta_{ll'}\sum_{j''l''}\phi_{\alpha\beta}\begin{pmatrix} jj'' \\ ll'' \end{pmatrix} \tag{6.16}$$

The second term in equation (6.16) arises from the interaction of any atom (jl) with the rest of the crystal, and is known as the *self term*.

In the general case, therefore, the equation of motion for the j-th atom in the l-th unit cell is given in this matrix formulation by:

$$m_j \ddot{\mathbf{u}}(jl,t) = -\sum_{j'l'} \mathbf{\Phi}\begin{pmatrix} jj' \\ ll' \end{pmatrix} \cdot \mathbf{u}(j'l',t) \tag{6.17}$$

where m_j is the mass of the j-th atom. We have now included the time (t) dependence in the displacement vector $\mathbf{u}(jl, t)$.

The solution for $\mathbf{u}(jl, t)$ will be a linear superposition of travelling harmonic waves of different wave vector $\mathbf{k}$ and mode label ν:

$$\mathbf{u}(jl,t)=\sum_{\mathbf{k},\nu}\mathbf{U}(j,\mathbf{k},\nu)\exp\left(i\left[\mathbf{k}\cdot\mathbf{r}(jl)-\omega(\mathbf{k},\nu)t\right]\right) \tag{6.18}$$

where $\mathbf{r}(jl)$ can be taken as either of two quantities. It can be taken as the equilibrium (or mean) position of the atom (jl), or else it can be taken as the origin of the unit cell (l). It is actually not important which choice is made, as the difference is simply transferred to the phase of $\mathbf{U}(j, \mathbf{k}, \nu)$, and the calculated mode frequencies are unaffected. The amplitude vector $\mathbf{U}(j, \mathbf{k}, \nu)$ (known as the *displacement vector*) is independent of l since the differences in motions between neighbouring unit cells are completely described by the exponential phase factor.

When we substitute the wave equation (6.18) into the equation of motion (6.17) we obtain the standard equation of motion:

$$m_j\omega^2(\mathbf{k},\nu)\mathbf{U}(j,\mathbf{k},\nu)=\sum_{j'l'}\Phi\begin{pmatrix}jj'\\0l'\end{pmatrix}\cdot\mathbf{U}(j',\mathbf{k},\nu)\exp\left(i\mathbf{k}\cdot\left[\mathbf{r}(j'l')-\mathbf{r}(j0)\right]\right) \tag{6.19}$$

where the reference atom is in the unit cell $l = 0$.

The equations of motion for a single solution (labelled ν) can now be expressed in vector form, where we quote the final result (compare with equation (6.2)):

$$\omega^2(\mathbf{k},\nu)\mathbf{e}(\mathbf{k},\nu)=\mathbf{D}(\mathbf{k})\cdot\mathbf{e}(\mathbf{k},\nu) \tag{6.20}$$

The column vector $\mathbf{e}(\mathbf{k}, \nu)$ is composed of the displacement vector weighted by the square root of the atomic mass, so that it has $3n$ elements (n is the number of atoms per unit cell):

$$\mathbf{e}(\mathbf{k},\nu)=\begin{pmatrix}\sqrt{m_1}U_x(1,\mathbf{k},\nu)\\ \sqrt{m_1}U_y(1,\mathbf{k},\nu)\\ \sqrt{m_1}U_z(1,\mathbf{k},\nu)\\ \sqrt{m_2}U_x(2,\mathbf{k},\nu)\\ \cdot\\ \cdot\\ \cdot\\ \sqrt{m_n}U_z(n,\mathbf{k},\nu)\end{pmatrix} \tag{6.21}$$

This is the generalisation of equation (6.1). $\mathbf{D}(\mathbf{k})$ is the $3n \times 3n$ dynamical matrix. Nomenclature becomes clumsy now! We write $\mathbf{D}(\mathbf{k})$ in terms of blocks of 3×3 matrices. Each block corresponds to pairs of atom labels j and j', and the elements of each block have labels α, $\beta = 1, 2, 3$, representing x, y, z respectively. The full matrix $\mathbf{D(k)}$ is composed of an $n \times n$ array of these smaller 3×3 matrices. The elements of the small 3×3 blocks of the dynamical matrix are given as:

$$D_{\alpha\beta}(jj',\mathbf{k}) = \frac{1}{\left(m_j m_{j'}\right)^{1/2}} \sum_{l'} \Phi_{\alpha\beta}\begin{pmatrix} jj' \\ 0l' \end{pmatrix} \exp\left(i\mathbf{k}\cdot\left[\mathbf{r}(j'l') - \mathbf{r}(j0)\right]\right) \quad (6.22)$$

where $l = 0$ as it refers to the reference unit cell (compare with the force constant matrix given above). The position of this element in the full dynamical matrix is clearly

$$3(j-1)+\alpha \;\; ; \;\; 3(j'-1)+\beta \quad (6.23)$$

This labelling can be seen to follow from the labelling in equations (6.19) and (6.21). It is straightforward to show that equation (6.3) is consistent with this definition.

As equation (6.20) has $3n$ components, there will be $3n$ solutions corresponding to the $3n$ branches in the dispersion diagram. We can compact our equations as before to make the $3n \times 3n$ matrix $\mathbf{e(k)}$, by joining together the column vectors $\mathbf{e}(\mathbf{k}, \nu)$, and now defining the frequency matrix $\Omega(\mathbf{k})$ as the diagonal matrix of the squares of the angular frequencies (compare with equation (6.6));

$$\Omega(\mathbf{k}) = \begin{pmatrix} \omega^2(\mathbf{k},1) & & & & & \\ & \omega^2(\mathbf{k},2) & & & & \\ & & \omega^2(\mathbf{k},3) & & & \\ & & & \cdot & & \\ & & & & \cdot & \\ & & & & & \omega^2(\mathbf{k},3n) \end{pmatrix} \quad (6.24)$$

$$\mathbf{e(k)}\cdot\Omega(\mathbf{k}) = \mathbf{D(k)}\cdot\mathbf{e(k)} \quad (6.25)$$

The dynamical matrix is *Hermitian*, i.e.

$$\mathbf{D(k)} = \left(\mathbf{D}^*(\mathbf{k})\right)^{\mathrm{T}} \quad (6.26)$$

We have already noted the property of Hermitian matrices that the eigenvalues are always real, and the eigenvectors, which may be complex, are orthogonal. As we pointed out earlier, the eigenvalues of the dynamical matrix are the squares of the angular frequencies of the different vibrational modes, and the eigenvectors give the relative displacements of the atoms associated with each vibrational mode.

The equations of motion do not contain information about the amplitude, which was discussed in detail in Chapter 4. The eigenvectors calculated by this procedure are normalised, such that:

$$\left(\mathbf{e}(\mathbf{k})\right)^{\mathrm{T}} \cdot \left(\mathbf{e}(\mathbf{k})\right)^{*} = \left(\mathbf{e}(\mathbf{k})\right)^{\mathrm{T}} \cdot \mathbf{e}(-\mathbf{k}) = 1 \tag{6.27}$$

The vector $\mathbf{e}(\mathbf{k}, \nu)$ is called the *polarisation vector*, and was first introduced in Chapter 4. Equation (6.27) is the generalisation of equation (6.8).

Let us stop to summarise what we have done. We have gone through this procedure in order to demonstrate how compact the equations for calculating the lattice vibration frequencies can be, and the form of the equations given here is now sufficiently general that they are readily incorporated into a computer program. In detail, we have simply redefined our variables in equation (6.21). The dynamical matrix includes both the force constant for any particular atom–atom interaction and the phase factor for the atomic motion of any wave. Thus the general equation (6.20) represents the equation of motion. The use of mass-weighted variables enables us to find solutions for ω^2 rather than $m\omega^2$. The formalism always gives real values for the solutions ω^2. Moreover, it generates the complete set of atomic motions associated with each wave. These motions are linearly independent (orthogonal), in that the motions associated with one wave do not generate the motions for any other. This can be expressed as (compare with equations (6.8) and (6.9)):

$$\left(\mathbf{e}(\mathbf{k}, \nu)\right)^{\mathrm{T}} \cdot \mathbf{e}(-\mathbf{k}, \nu') = \delta_{\nu\nu'} \tag{6.28}$$

These are called *normal modes*, and are the fundamental vibrational motions. Although the frequencies are real, the motions are in general complex, which simply expresses through the real and imaginary parts the relative phases of the motions of each atom.

Extension for molecular crystals

The theory we have outlined in this chapter can also be extended for molecular crystals, where there are also rotational degrees of freedom. Each molecule has six dynamical variables rather than three. It is easier to define a coordinate system for each molecule such that the tensor for the moment of inertia is diagonal, although for all but the highest-symmetry cases the coordinate system for one molecule will not be coincidental with that for other molecules or with the crystal coordinate system. In the components of the mode eigenvectors involving the rotational degrees of freedom, and in the terms in the dynamical matrix involving the differentials with respect to rotations, the factors of $\sqrt{m}$ will be replaced by $\sqrt{I_{jj}}$, where I_{jj} is the relevant diagonal component of the inertia tensor. The theory of the lattice dynamics of molecular crystals was originally developed by Cochran and Pawley (1964) and Pawley (1967, 1972) for rigid molecules, and has since been adapted to include internal degrees of freedom (Chaplot et al. 1982, 1983).

The dynamical matrix and symmetry

We remarked in Chapter 3 that the normal modes can be described by symmetry, in that the displacements generated by the normal mode will transform as a given representation of the symmetry of the wave vector. The dynamical matrix formalism can be applied directly to the determination of the symmetry of the normal modes, and this approach lends itself to computer programming, enabling the symmetries of normal modes to be calculated by computer rather than hand. The essential theory is described by Maradudin and Vosko (1968) and Warren (1968), and the actual implementation of this approach is outlined by Warren and Worton (1974). The dynamical matrix is solved for modes that involve displacements of one atom type only (by which we mean the set of atoms that are related by the space group of the crystal). The calculated eigenvectors are then grouped according to their symmetry. The actual mode eigenvectors that are obtained from a lattice dynamics calculation will be a simple linear combination of the symmetry-adapted eigenvectors, so the symmetry of the normal modes can be assigned following the use of a simple routine to solve the corresponding set of simultaneous equations. The group theory computer program of Warren and Worton (1974) allows for the use of both atoms and molecules.

Extension for the shell model

The formalism presented so far has been appropriate for the case where all the particles are atoms with mass. This is the so-called *rigid-ion* model, in which it is assumed that the ions cannot deform. The alternative to the rigid-ion model is the shell model, which was described in Chapter 1. The shell model is a simple method for modelling the deformation of the electronic structure of an ion due to the interactions with other atoms, notably associated with the polarisation induced by local electric fields. It is straightforward to modify the dynamical matrix to take account of shell-model interactions. The shell model effectively splits an atom into two charged particles: a massless shell and a core that contains all the mass of the atom. We therefore have equations of motion for both core and shell, although we will now show that this does not increase the number of degrees of freedom of the system nor the number of normal modes. The essential point is that the zero mass of the shell forces the motion of the shell to be determined by the motion of the core. This is associated with the fact that the shell is always in an equilibrium position, since its zero mass means that it can move to a new position instantaneously. This is the adiabatic approximation that was introduced in Chapter 1.

We can rewrite the equations of motion (6.17) for the cores and shells:

$$m_j \ddot{\mathbf{u}}_c(jl,t) = -\sum_{j'l'} \mathbf{\Phi}_{cc} \cdot \mathbf{u}_c(j'l',t) - \sum_{j'l'} \mathbf{\Phi}_{cs} \cdot \mathbf{u}_s(j'l',t) \tag{6.29}$$

$$0 = -\sum_{j'l'} \mathbf{\Phi}_{sc} \cdot \mathbf{u}_c(j'l',t) - \sum_{j'l'} \mathbf{\Phi}_{ss} \cdot \mathbf{u}_s(j'l',t) \tag{6.30}$$

where the subscripts c and s indicate core and shell respectively. For the force constants, the subscripts indicate that the derivatives are with respect to the displacements of the core or shell; terms with both core and shell are of course allowed. By analogy with equation (6.20) we can write the general equations of motion for both cores and shells as:

$$\omega^2(\mathbf{k}, \nu)\mathbf{e}_c = \mathbf{D}_{cc} \cdot \mathbf{e}_c + \mathbf{D}_{cs} \cdot \mathbf{e}_s \tag{6.31}$$

$$0 = \mathbf{D}_{sc} \cdot \mathbf{e}_c + \mathbf{D}_{ss} \cdot \mathbf{e}_s = \mathbf{D}_{cs}^{+} \cdot \mathbf{e}_c + \mathbf{D}_{ss} \cdot \mathbf{e}_s \tag{6.32}$$

The eigenvectors $\mathbf{e}_c$ retain the factors of $m^{1/2}$ as in equation (6.21), but owing to the mass of the shells being zero this is not the case for $\mathbf{e}_s$. Accordingly, whereas the core–core dynamical matrix $\mathbf{D}_{cc}$ contains the weighting $1/(m_j m_{j'})^{1/2}$, $\mathbf{D}_{cs}$ is weighted only by $1/m^{1/2}$ and there is no mass weighting of $\mathbf{D}_{ss}$. Equation (6.32) gives us the equation that relates the displacements of the shells to the cores, confirming that the core–shell model does not have more degrees of freedom than a rigid-ion model:

$$\mathbf{e}_s = -\mathbf{D}_{ss}^{-1} \cdot \mathbf{D}_{cs}^{+} \cdot \mathbf{e}_c \tag{6.33}$$

We substitute equation (6.33) into equation (6.31) to give the final equation of motion:

$$\omega^2(\mathbf{k}, \nu)\mathbf{e}_c = \left[\mathbf{D}_{cc} - \mathbf{D}_{cs} \cdot \mathbf{D}_{ss}^{-1} \cdot \mathbf{D}_{cs}^{+}\right] \cdot \mathbf{e}_c \tag{6.34}$$

This modified dynamical equation can be solved for the dispersion curves and mode eigenvectors in the same way that equation (6.20) and subsequent generalisation is solved for the rigid-ion case.

The treatment we have outlined is quite general, and, for example, it does not assume that the positions of the shells are the same as the positions of the corresponding cores. Thus an atom that does not occupy a centre of symmetry in a crystal structure can be polarised at equilibrium, and this is usually the case when a structure is relaxed in a lattice energy calculation with a realistic interatomic potential model. In the early development of the shell model it was assumed that the core and shell positions are coincident. This allowed equation (6.34) to be written in a different form, which we give here for reference. The core–core term can be written as

$$\mathbf{D}_{cc} = \mathbf{R} + \mathbf{Z} \cdot \mathbf{C} \cdot \mathbf{Z} \tag{6.35}$$

The matrix **R** contains the short-range interactions between cores, the matrix **Z** is a diagonal matrix with each element being the charge of the core, and the matrix **C** is the so-called "Coulomb matrix", which expresses the long-range part of the dynamical matrix. Analogous equations follow for the shell–shell term and the core–shell term:

$$\mathbf{D}_{ss} = \mathbf{S} + \mathbf{Y} \cdot \mathbf{C} \cdot \mathbf{Y} \tag{6.36}$$

$$\mathbf{D}_{cs} = \mathbf{T} + \mathbf{Z} \cdot \mathbf{C} \cdot \mathbf{Y} \tag{6.37}$$

where the diagonal matrix **Y** contains the charges of the shells, **S** contains the short-range interactions between the shells, and **T** contains the short-range interactions between the cores and shells. The final dynamical matrix is therefore given as

$$\mathbf{D} = \mathbf{R} + \mathbf{Z} \cdot \mathbf{C} \cdot \mathbf{Z} + (\mathbf{T} + \mathbf{Z} \cdot \mathbf{C} \cdot \mathbf{Y}) \cdot (\mathbf{S} + \mathbf{Y} \cdot \mathbf{C} \cdot \mathbf{Y})^{-1} \cdot \left(\mathbf{T}^{+} + \mathbf{Y} \cdot \mathbf{C} \cdot \mathbf{Z}\right) \tag{6.38}$$

An alternative parameterisation of equation (6.38) has also been used (Cochran 1971). Further developments of this basic shell model are described in Bilz and Kress (1979, ch. 2).

The shell model has also been extended for molecular crystals by Luty and Pawley (1974, 1975) and Pawley and Leech (1977). In the simplest application it is assumed that the whole molecule has a uniform polarisability. The early work was applied to the lattice dynamics of sulphur (Luty and Pawley 1975), but there have been relatively few other applications.

Actual calculations of dispersion curves

The formalism we have introduced may seem a little formidable, but computer programs are available that can calculate the dispersion curves very quickly using standard routines to diagonalise the dynamical matrix. These calculations are often used in support of experimental studies, the methods of which are outlined in Chapter 9. There are two approaches in the analysis of measured dispersion curves, which may both involve fitting calculated curves to the experimental ones.

The first approach makes no assumption about the form of the interatomic potentials, other than the Coulombic interaction. Instead the values of the relevant set of harmonic force constants are simply treated as numbers that are obtained by fitting calculated dispersion curves to measurements. If central forces (i.e. forces that depend only on separation distances and not on the direction of the interatomic separation) are used, there are two force constants associated with each pair of atoms, namely the radial force constant K_r that acts along the separation distance, and the tangential force constant K_t that acts perpendicularly to the separation distance. These are both related to the short range interatomic potential φ_R by:

$$K_r = \frac{\partial^2 \varphi}{\partial r^2} \qquad (6.39)$$

$$K_t = \frac{1}{r}\frac{\partial \varphi}{\partial r} \qquad (6.40)$$

The first derivatives of φ_R are subject to the equilibrium constraint that any derivative of the lattice energy must be zero (equation (1.11)). With the use of a shell model, the force constants are usually required only for close neighbours in order to give a good fit to experimental dispersion curves. Moreover it can often be assumed that the short-range forces operate only between the shells, so that the matrices **R** and **T** can be set to zero. Further levels of approximation include allowing only the shells to be charged (**Z** = 0), and to use the core–shell representation only for negative ions.

The force constant approach was first used for ionic crystals following the development of neutron scattering methods and the first measurements of dispersion curves (Woods et al. 1960, 1963); it is also used for metals and semiconductors (e.g. Cochran 1959c), where the forces are not easily described by mathematical functions. However, this approach becomes more difficult when the symmetry is low (the number of independent force constants can be reduced significantly by high symmetry), the number of atoms is high, and when forces are long-ranged – in each case a large number of force constants is then required.

The second approach is to use a model interatomic potential energy function, which will contain a few variables with values that can be determined by fitting to the measured dispersion curves. The advantage of this approach is that the given model provides a constraint that relates the values of the individual force constants. For example, for a given pair of atoms the two force constants K_r and K_t are no longer independent. It is usual that the model is subject to the additional constraint that the predicted equilibrium crystal structure should be close to the experimental structure. In general the calculations of the dispersion curves should be performed using an equilibrium structure for the model rather than the observed structure, so that the tangential force constants (equation (6.40)) do not include non-equilibrium contributions. This approach works best for insulators, where the Coulombic, dispersive and repulsive interactions can be modelled easily; the variables might relate to the size and hardness of the atoms (Chapter 1).

The force constant models can usually be made to give a better representation of the dispersion curves than the model interatomic potentials, owing to the fewer number of constraints. Interatomic potentials are also subject to correlations between parameters, which further reduces the number of degrees of freedom in the model.

As an example of the comparison between the two approaches we consider the case of calcite. Measured dispersion curves are given in Figure 3.8. The measured dispersion curves were originally fitted by a force constant shell model, which gave good agreement between the calculated and measured frequencies (Cowley and Pant 1973). More recently the dispersion curves have been calculated using an interatomic potential model (using rigid ions) that was partly optimised by fitting to the elastic constants and frequencies at $\mathbf{k} = 0$ (Dove et al. 1992c). The agreement between the calculated and measured dispersion curves is poorer than for the force constant model, but is nevertheless reasonable (the calculated dispersion curves using this model are also given in Figure 3.8). However, the principal advantage of the interatomic potential model is that it can be applied to other problems, whereas the force constant model can be used only for calculations of dispersion curves. The model for calcite has also been used to calculate the energies of different phases of calcite, and in further studies of the structural properties (Dove et al. 1992c). In this case the dispersion curves were primarily used to optimise the model, an illustration of the importance of such measurements.

There are of course many cases when we already have a reasonable interatomic potential model and do not need to optimise against experimental data.

This is particularly true for classes of systems for which we have transferable potentials, such as organic crystals and silicates (Chapter 1). In such cases dispersion curves may be required for further calculations of thermodynamic properties, for example, and having a good model means that expensive or difficult (even impossible) experimental measurements of dispersion curves are not required. In other cases, a good model calculation can be used to help interpret measurements of dispersion curves.

A recent description of a computer program for lattice dynamics calculations is given by Eckold et al. (1987). This program includes the group theory program of Warren and Worton (1974), and routines for calculating intensities for neutron scattering, using the formalism of Chapter 9, and for fitting force constants to measured dispersion curves.

Normal mode coordinates

In Chapter 4 we introduced the normal mode coordinate, $Q(\mathbf{k}, \nu)$, such that:

$$\mathbf{u}(jl,t) = \frac{1}{\left(Nm_j\right)^{1/2}} \sum_{\mathbf{k},\nu} \mathbf{e}(j,\mathbf{k},\nu)\exp\left(i\mathbf{k}\cdot\mathbf{r}(jl)\right)Q(\mathbf{k},\nu) \tag{6.41}$$

$$\dot{\mathbf{u}}(jl,t) = \frac{1}{\left(Nm_j\right)^{1/2}} \sum_{\mathbf{k},\nu} \mathbf{e}(j,\mathbf{k},\nu)\exp\left(i\mathbf{k}\cdot\mathbf{r}(jl)\right)\dot{Q}(\mathbf{k},\nu) \tag{6.42}$$

The dynamic energy of the harmonic system, called the *Hamiltonian* $\mathcal{H}$, can be written as

$$\mathcal{H} = \frac{1}{2}\sum_{jl} m_j \left|\dot{\mathbf{u}}(jl)\right|^2 + \frac{1}{2}\sum_{jj',ll'} \mathbf{u}^{\mathrm{T}}(jl)\cdot\mathbf{\Phi}\begin{pmatrix} jj' \\ ll' \end{pmatrix}\cdot\mathbf{u}(j'l') \tag{6.43}$$

This equation follows from equation (6.11). We can substitute for $\dot{\mathbf{u}}$ and $\mathbf{u}$ in equation (6.43). It turns out that when we do this, we end up with the satisfying result introduced in equation (4.6):

$$\mathcal{H} = \frac{1}{2}\sum_{\mathbf{k},\nu} \dot{Q}(\mathbf{k},\nu)\dot{Q}(-\mathbf{k},\nu) + \frac{1}{2}\sum_{\mathbf{k},\nu} \omega^2(\mathbf{k},\nu)Q(\mathbf{k},\nu)Q(-\mathbf{k},\nu) \tag{6.44}$$

The first part of equation (6.44) was first introduced in Chapter 4, and is derived in Appendix B (equation (B.6)). The second part of equation (6.44), the potential energy term, is also derived in Appendix B. The derivation uses the dynamical matrix transformation that is given by equations (6.17)–(6.25);

the potential energy term in equation (6.44) is clearly a diagonalisation of the same term in equation (6.43).

Summary

In this chapter we have bridged the gap between the simple models of earlier chapters and real systems.

1. We have obtained general equations for the harmonic lattice vibrations that can be used for any crystal.
2. We have defined the dynamical matrix, noting that the vibrational frequencies are obtained as the square roots of the eigenvalues and the corresponding atomic motions are given by the eigenvectors of the dynamical matrix.
3. The formalism has been extended to include shell-model interactions.
4. The formalism for the mode eigenvectors has been recast in terms of the normal mode coordinates, and the harmonic energy of the crystal has been written in terms of these new variables.

FURTHER READING

Ashcroft and Mermin (1976) ch. 22
Born and Huang (1954) ch. 15,38
Brüesch (1982) ch. 2, 3, 4; app. F–J
Califano et al. (1981)
Cochran (1973) ch. 4
Willis and Pryor (1975) ch. 3

7

Acoustic modes and macroscopic elasticity

The connection between the acoustic modes and the elastic properties of a crystal is developed. The dynamical matrix for the acoustic modes is written in terms of the elastic constant tensor.

The behaviour of long-wavelength acoustic modes

We pointed out in Chapter 2 that the acoustic modes in the long-wavelength limit correspond to crystal strains, and that the force constants that determine the dispersion curves are given by the appropriate elastic constants. The purpose of this chapter is to develop the relationship between the acoustic modes and the complete elastic constant tensor.

Consider first a longitudinal acoustic mode in a cubic crystal with wave vector along [100], as illustrated in Figure 7.1. The magnitude of the wave vector is small but non-zero, such that the wavelength is much larger than the unit cell size. Each (100) plane of atoms is displaced in the x direction by a constant amount relative to its neighbouring planes. Therefore the displacement u_x of each plane is proportional to its position x. This corresponds to a uniform compressional strain of the crystal, $e_{11} = \partial u_x / \partial x$, which locally makes a cubic unit cell tetragonal.

Now consider a transverse acoustic mode in a cubic crystal with wave vector along [100], as illustrated in Figure 7.2. In this case the planes of atoms are displaced along the y direction by a constant amount relative to its neighbouring planes. Therefore the displacement u_y of each plane is also proportional to its position x and $\partial u_y / \partial x$ is constant. Because the direction of the displacements u_y is orthogonal to the x direction, the gradient $\partial u_y / \partial x$ describes the shear strain e_{12}. This strain will locally make an orthogonal unit cell monoclinic.

Finally, consider the transverse acoustic mode in a cubic crystal with wave vector along [110] illustrated in Figure 7.3. The planes of atoms are displaced

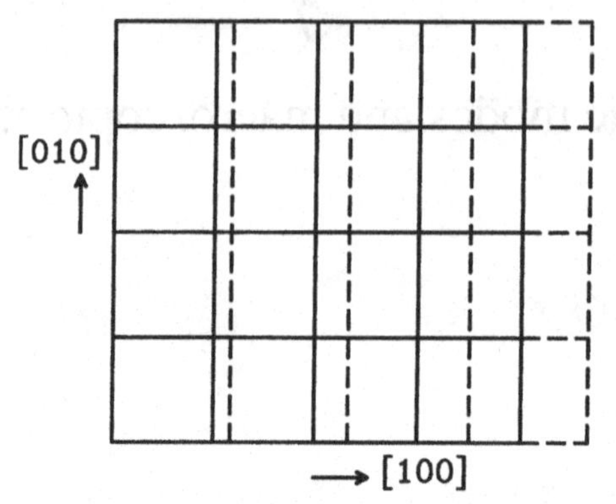

Figure 7.1: Atomic displacements associated with a long-wavelength longitudinal acoustic mode propagating along [100] in a cubic crystal.

along [1$\bar{1}$0]. In this case the displacements that accompany the shear strain give rise to a local tetragonal distortion of the unit cell plus a rotation of the unit cell. There is also a subsidiary shear of the unit cell that makes the axes non-orthogonal, but as this distortion is of higher order than the tetragonal shear strain it is a much smaller effect and can be neglected.

Acoustic mode frequencies and the elastic constant tensor

In this section we present the method for calculating the slopes of the acoustic phonon dispersion curves in the long-wavelength limit, where the acoustic modes give rise to strain distortions as described above. The pure strain component ε_{ij} is classically defined as

$$\varepsilon_{ij} = \frac{1}{2}\left(\frac{\partial u_i}{\partial r_j} + \frac{\partial u_j}{\partial u_i}\right) = \frac{1}{2}\left(e_{ij} + e_{ji}\right) \tag{7.1}$$

where $\mathbf{u} = (u_1, u_2, u_3)$ is the displacement of a volume element caused by the strain at the position $\mathbf{r} = (r_1, r_2, r_3)$, and the subscripts i and j denote the components of the vectors $\mathbf{u}$ and $\mathbf{r}$. The energy per unit volume U associated with a set of strain distortions is given as[1]

$$U = \frac{1}{2}\sum_{i,j,k,l} C_{ijkl}\varepsilon_{ij}\varepsilon_{kl}^{*} \tag{7.2}$$

[1] We have to take the complex conjugate of the second strain component in order to ensure that the energy is a real quantity. This would not be necessary if we did not use a complex notation for the travelling waves.

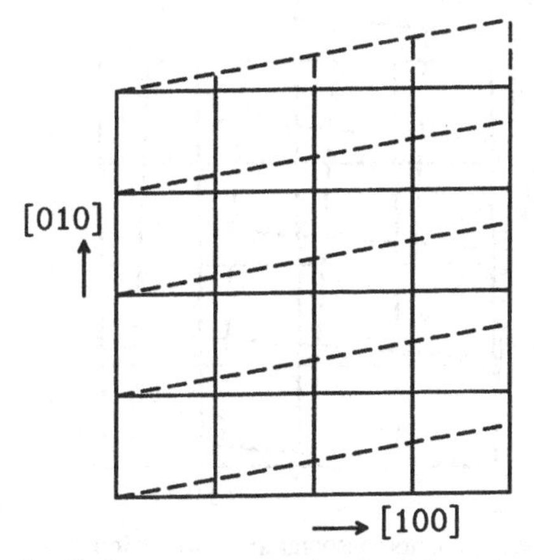

Figure 7.2: Atomic displacements associated with a long-wavelength transverse acoustic mode propagating along [100] in a cubic crystal and polarised along [010].

where **C** is the elastic constant tensor, which is of fourth rank. The energy of an element of volume dV due to the strains is therefore given as[2]

$$\mathrm{d}E = U\mathrm{d}V = \frac{\mathrm{d}V}{2}\sum_{i,j,k,l} C_{ijkl}\varepsilon_{ij}\varepsilon_{kl}^{*} \tag{7.3}$$

In the present context, the strains arise from the acoustic modes, which propagate as

$$\mathbf{u}(t) = \tilde{\mathbf{u}}\exp(i[\mathbf{k}\cdot\mathbf{r} - \omega t]) \tag{7.4}$$

where $\mathbf{k} = (k_1, k_2, k_3)$. Noting the definitions (7.1) and (7.4) we see that the strain waves propagate as

$$\begin{aligned}\varepsilon_{ij}(t) &= \frac{i}{2}\left(u_i(t)k_j + u_j(t)k_i\right)\\ &= \frac{i}{2}\left(\tilde{u}_i k_j + \tilde{u}_j k_i\right)\exp(i[\mathbf{k}\cdot\mathbf{r} - \omega t])\end{aligned} \tag{7.5}$$

To calculate the dynamics of the small volume element we use the standard Newton equation of motion for the mass of the volume element to obtain:

$$\rho\ddot{u}_j\mathrm{d}V = -\frac{\partial}{\partial u_j}\mathrm{d}E \tag{7.6}$$

where ρ is the density.

[2] We have assumed that the diameter of the volume element is much smaller than the wavelength of the acoustic modes we will be considering.

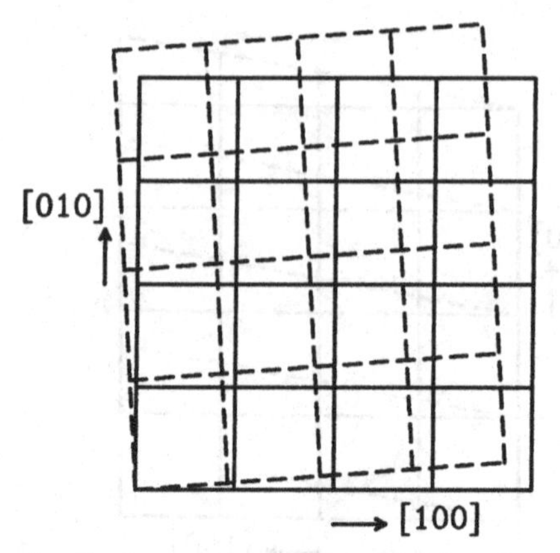

Figure 7.3: Atomic displacements associated with a long-wavelength transverse acoustic mode propagating along [110] in a cubic crystal and polarised along $[1\bar{1}0]$.

We can solve equation (7.6) in two steps. Firstly, from equation (7.4) we have:

$$\rho \ddot{u}_i \mathrm{d}V = -\rho\omega^2 u_i \mathrm{d}V \tag{7.7}$$

Secondly, from equation (7.3) and the first line of equation (7.5) we have:

$$\begin{aligned}
-\frac{\partial}{\partial u_i}\mathrm{d}E &= -\frac{\mathrm{d}V}{2}\sum_{j,k,l} C_{ijkl}\frac{\partial \varepsilon_{ij}}{\partial u_i}\varepsilon_{kl}^* \\
&= -\frac{\mathrm{d}V}{2}\sum_{j,k,l} C_{ijkl}k_j\left(u_k k_l + u_l k_k\right) \\
&= -\mathrm{d}V\sum_{j,k,l} C_{ijkl}k_j k_l u_k
\end{aligned} \tag{7.8}$$

Gathering together these last two equations leads to the result:

$$\rho\omega^2 \tilde{u}_i = \sum_{j,k,l} C_{ijkl}k_j k_k \tilde{u}_l \tag{7.9}$$

which can be written in matrix form:

$$\rho\omega^2 \tilde{\mathbf{u}} = \mathbf{M}\cdot\tilde{\mathbf{u}} \tag{7.10}$$

where the dynamical matrix $\mathbf{M}$ is given as

$$M_{ik} = \sum_{j,l} C_{ijkl}k_j k_l \tag{7.11}$$

From equation (7.10) we see that $\rho\omega^2$ is the eigenvalue of the matrix **M** (obtained as one of the solutions of the determinant of **M**), and the eigenvectors give the corresponding motions. **M** is therefore identical to the dynamical matrix for the acoustic modes.

The reader is refered to Nye (1964, ch. VIII) for a discussion of the symmetry properties of the elastic constant tensor. There is a general reduction of the number of *independent* components from 81 to 21 in the least symmetric case (triclinic), and in the most symmetric case (cubic) there are only 3 independent components. It is common practice to use the Voigt notation, in which pairs of indices are replaced by single indices:

$$\begin{array}{ccc} 11 \rightarrow 1 & 22 \rightarrow 2 & 33 \rightarrow 3 \\ 12,21 \rightarrow 6 & 13,31 \rightarrow 5 & 23,32 \rightarrow 4 \end{array}$$

The elastic constant tensor can then be represented by a symmetric 6×6 matrix. Nye (1964) tabulates the elastic constant tensor in this representation for each crystal symmetry.

One application of the formalism presented here is that it is possible to obtain elastic constants from measurements of the acoustic mode dispersion relations.

Worked example: acoustic waves in a cubic crystal

The elastic constant matrix for a cubic crystal is given as

$$\mathbf{C} = \begin{pmatrix} C_{11} & C_{12} & C_{12} & \cdot & \cdot & \cdot \\ C_{12} & C_{11} & C_{12} & \cdot & \cdot & \cdot \\ C_{12} & C_{12} & C_{11} & \cdot & \cdot & \cdot \\ \cdot & \cdot & \cdot & C_{44} & \cdot & \cdot \\ \cdot & \cdot & \cdot & \cdot & C_{44} & \cdot \\ \cdot & \cdot & \cdot & \cdot & \cdot & C_{44} \end{pmatrix} \tag{7.12}$$

where the dots indicate zero values. The symmetric dynamical matrix **M** then has the form (showing only the upper right half):

$$\mathbf{M} = \begin{pmatrix} C_{11}k_1^2 + C_{44}\left(k_2^2 + k_3^2\right) & \left(C_{12} + C_{44}\right)k_1k_2 & \left(C_{12} + C_{44}\right)k_1k_3 \\ & C_{11}k_2^2 + C_{44}\left(k_1^2 + k_3^2\right) & \left(C_{12} + C_{44}\right)k_2k_3 \\ & & C_{11}k_3^2 + C_{44}\left(k_1^2 + k_2^2\right) \end{pmatrix} \tag{7.13}$$

The solutions for the three wave vectors, $[\xi, 0, 0]$, $[\xi, \xi, 0]$ and $[\xi, \xi, \xi]$, are tabulated below:

$\mathbf{k} = [\xi, 0, 0]$	$\mathbf{k} = [\xi, \xi, 0]$	$\mathbf{k} = [\xi, \xi, \xi]$
$\rho\omega^2_{[100]} = C_{11}\xi^2$	$\rho\omega^2_{[110]} = (C_{11} + C_{12} + 2C_{44})\xi^2$	$\rho\omega^2_{[111]} = (C_{11} + 2C_{12} + 4C_{44})\xi^2$
$\rho\omega^2_{[010]} = C_{44}\xi^2$	$\rho\omega^2_{[1\bar{1}0]} = (C_{11} - C_{12})\xi^2$	$\rho\omega^2_{[1\bar{1}0]} = (C_{11} - C_{12} + C_{44})\xi^2$
$\rho\omega^2_{[001]} = C_{44}\xi^2$	$\rho\omega^2_{[001]} = 2C_{44}\xi^2$	$\rho\omega^2_{[11\bar{2}]} = (C_{11} - C_{12} + C_{44})\xi^2$

where the subscripts on ω denote the directions of motion of the atoms.

We can comment on the stability of the crystal. If C_{44} is negative, the crystal is unstable against the shear given by one of the transverse acoustic modes with wave vectors in the **a*–b*** plane (and related planes). Alternatively, if $C_{11} \leq C_{12}$ the crystal is unstable against the transverse acoustic mode with the wave vector along [110] and polarisation vector along $[1\bar{1}0]$. There are a number of phase transitions that occur in all crystal classes when these stability conditions are broken on cooling, as discussed in Chapter 8. These phase transitions are called *ferroelastic*; an extensive review of the properties of ferroelastic phase transitions is given by Salje (1990). The stability conditions for the acoustic modes for all crystal classes have been enumerated by Cowley (1976) and Terhune et al. (1985).

Summary

1 We have shown that the long-wavelength acoustic modes generate shear distortions of the crystal structure.
2 We have derived the dynamical matrix for the acoustic modes, which is constructed in the general case in terms of the elastic constants.
3 Using the acoustic dynamical matrix it is possible to obtain values for the elastic constants from the acoustic mode dispersion curves.

FURTHER READING

Ashcroft and Mermin (1976) ch. 22
Born and Huang (1954) ch. 11–13, 26–27
Brüesch (1982) pp 84–92
Nye (1964) ch. VIII

8
Anharmonic effects and phase transitions

The effects of anharmonic interactions are described in detail. Most of the chapter is concerned with the role of anharmonic interactions as the driving mechanism of displacive phase transitions. The soft mode model of phase transitions is developed and discussed in connection with various different kinds of phase transitions.

Failures of the harmonic approximation

The harmonic phonon model developed in the previous chapters has given us many reasonable results (phonon frequencies, normal mode amplitudes, mean-squared atomic displacements, elastic constants etc.). However, there are a number of phenomena that cannot be explained within the harmonic approximation. These are of three types:

1 Temperature dependence of equilibrium properties:
  - thermal expansion;
  - temperature dependence of elastic constants;
  - temperature dependence of phonon frequencies;
  - natural linewidths of phonon frequencies.
2 Occurrence of phase transitions.
3 Transport properties, e.g. thermal conductivity.

These effects result from the anharmonic interactions. The aim of this chapter is to explore the effects of anharmonic interactions.

Anharmonic interactions

We can start by taking the harmonic Hamiltonian given in terms of normal mode coordinates, equation (6.31), and then simply add to this the corresponding anharmonic terms:

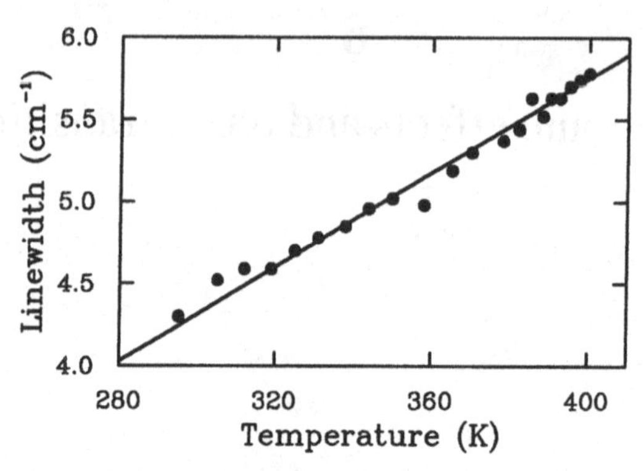

Figure 8.1: Linewidth of a $\mathbf{k} \sim 0$ optic mode in KNO_3 as a function of temperature. The straight line has been fitted to the data, and passes though zero width at $T = 0$ K (Harris 1992).

$$\mathcal{H} = \frac{1}{2}\sum_{\mathbf{k},\nu} \dot{Q}(\mathbf{k},\nu)\dot{Q}(-\mathbf{k},\nu) + \frac{1}{2}\sum_{\mathbf{k},\nu} \omega_0^2(\mathbf{k},\nu) Q(\mathbf{k},\nu) Q(-\mathbf{k},\nu)$$

$$+\sum_{\kappa>2}\frac{1}{\kappa!}\sum_{\mathbf{k}_1,\nu_1}\cdots\sum_{\mathbf{k}_\kappa,\nu_\kappa} V_\kappa\begin{pmatrix}\mathbf{k}_1\cdots\mathbf{k}_\kappa\\ \nu_1\cdots\nu_\kappa\end{pmatrix} Q(\mathbf{k}_1,\nu_1)\cdots Q(\mathbf{k}_\kappa,\nu_\kappa)\Delta(\mathbf{k}_1+\cdots+\mathbf{k}_\kappa) \tag{8.1}$$

where V_κ is the κ-th order coupling constant, which gives the strength of the interaction between the relevant phonons, $\omega_0(\mathbf{k}, \nu)$ is the harmonic frequency of the mode $(\mathbf{k}, \nu)$, and the function $\Delta(\mathbf{G})$ has a value of unity if $\mathbf{G}$ is a reciprocal lattice vector and zero otherwise.

We shall continue using the approximation of small oscillations so that we need only consider the cubic and quartic terms. The anharmonic interactions have two main effects on the phonons. Firstly, they change the phonon frequencies from the harmonic values; we will consider this in some detail later in this chapter. Secondly, they cause the phonon modes to dampen. This effect is seen as a broadening of the spectral lines in a scattering experiment. Figure 8.1 shows a phonon linewidth, measured in an infrared absorption experiment, that increases linearly with temperature. A spectral linewidth (units of frequency) is the inverse of a lifetime of an excited state. In this case the phonon represents an excited state of the crystal, which after a period of time (the lifetime) decays into another state. One can think of phonons scattering from one another as atoms in a gas scatter from each other. The cubic and quartic terms represent the scattering processes shown in Figure 8.2. There are strict conservation laws that restrict the behaviour of phonon scattering processes, which we illustrate with respect to three-phonon processes. The first is the conservation

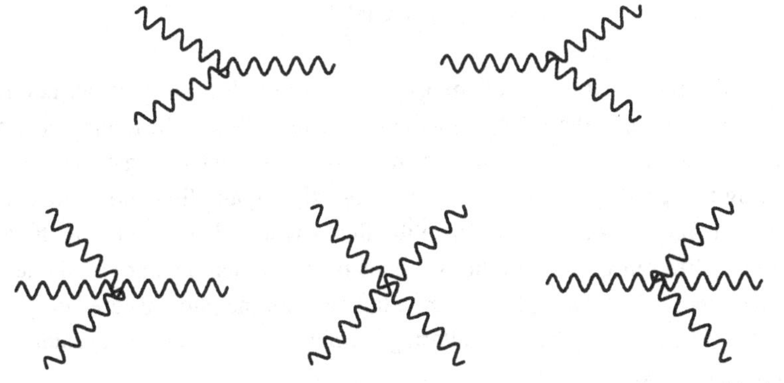

Figure 8.2: Schematic representations of phonon collisions corresponding to cubic (top) and quartic (bottom) anharmonic interactions.

of energy. For two phonons combining to form a third this is given by the frequency relationship:

$$\omega_1 + \omega_2 = \omega_3 \tag{8.2}$$

The second is the conservation of wave vector. This has the property that the total wave vector can change only by a reciprocal lattice vector **G**, as for example,

$$\mathbf{k}_1 + \mathbf{k}_2 = \mathbf{k}_3 + \mathbf{G} \tag{8.3}$$

This is like any scattering process (e.g. Bragg scattering of X-rays). The case of $\mathbf{G} = 0$ is called a normal scattering process (exact conservation of wave vector); for non-zero values of **G** the process is called an *Umklapp process* after the German for *flipping over*. For this reason the number of terms in the anharmonic Hamiltonian are restricted to those cases that obey these conservation laws. One other condition is given by symmetry: the Hamiltonian can contain only terms that conserve the symmetry of the system. The symmetry of a normal mode coordinate is an irreducible representation of the space group of the crystal, whereas the Hamiltonian must have the full symmetry of the space group. Therefore the terms in the Hamiltonian must contain products of the irreducible representations that give the identity representation.

Simple treatment of thermal conductivity

The simple theory of thermal conductivity highlights nicely the anharmonic phonon scattering processes. The thermal conductivity of an insulator along a direction x can be represented by the equation:

$$J = -K\frac{\mathrm{d}T}{\mathrm{d}x} \tag{8.4}$$

where J is the amount of heat energy passing through a unit area per unit time (the heat flux), K is the thermal conductivity, and $\mathrm{d}T/\mathrm{d}x$ is the temperature gradient. If the crystal were harmonic, we could pump energy into one end (by heating that end only) and generate phonons that would flow along the crystal until reaching the other end. Hence the flux J would depend only on the temperature difference between the two ends of the crystal, and not on the length of the crystal. That this is not the case tells us that the phonons are being scattered. There are a number of scattering mechanisms; the one we consider here is *phonon–phonon scattering*.[1]

We can sketch a rough derivation for the thermal transport equation given above and obtain an expression for the thermal conductivity K using an analogy with the kinetic theory of gases, considering the phonons to be interacting particles. We consider two small regions of a crystal separated by a phonon mean-free-path length λ. When there is a temperature gradient $\mathrm{d}T/\mathrm{d}x$ through the crystal (parallel to the distance between our two regions, defined as the x direction), the temperature difference over this mean-free-path length is

$$\Delta T = \lambda_x \frac{\mathrm{d}T}{\mathrm{d}x} = v_x \tau \frac{\mathrm{d}T}{\mathrm{d}x} \tag{8.5}$$

where v_x is the x-component of the average phonon drift velocity, and τ is the mean phonon lifetime. The difference between the phonon densities in our two regions is simply given as

$$n(T+\Delta T) - n(T) = \Delta T \frac{\mathrm{d}n}{\mathrm{d}T} = v_x \tau \frac{\mathrm{d}T}{\mathrm{d}x}\frac{\mathrm{d}n}{\mathrm{d}T} \tag{8.6}$$

The energy flux J is simply given as the product of the average drift velocity, the phonon energies, and the difference between the occupation numbers:[2]

$$\begin{aligned} J &= -v_x \sum_j \hbar\omega_j \left[n_j(T+\Delta T) - n_j(T) \right] = -v_x^2 \tau \frac{\mathrm{d}T}{\mathrm{d}x} \sum_j \hbar\omega_j \frac{\mathrm{d}n_j}{\mathrm{d}T} \\ &= -v_x^2 \tau C_V \frac{\mathrm{d}T}{\mathrm{d}x} = -\frac{1}{3} v^2 \tau C_V \frac{\mathrm{d}T}{\mathrm{d}x} = -K\frac{\mathrm{d}T}{\mathrm{d}x} \end{aligned} \tag{8.7}$$

[1] Other mechanisms include scattering by defects (which we will not go into here because we have not considered the effects of defects on the phonon spectrum), scattering from the surfaces (which we can neglect if we consider thick samples) and scattering of phonons by electrons, the so-called *electron–phonon interaction*, which is the most important process in metals but not so important in insulators.

[2] We assume for simplicity that there is no correlation between the average phonon drift velocity or phonon lifetime and the particular mode.

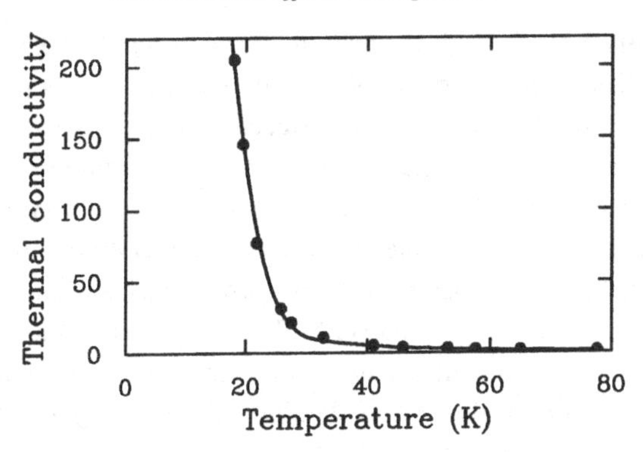

Figure 8.3: Thermal conductivity of a crystal of NaF (Jackson and Walker 1971).

where C_V is the heat capacity (considered here to be at constant volume). Note that in equation (8.7) we have replaced the average of the squared velocity component by the average of the squared total velocity. Hence we have obtained the simple result:

$$K = \frac{1}{3} v^2 \tau C_V = \frac{1}{3} v \lambda C_V \tag{8.8}$$

In Figure 8.3 the thermal conductivity of NaF is shown as a function of temperature. As expected, it falls on increasing temperature because the number of collisions increases so that the mean-free-path length decreases. This effect dominates the increase in the heat capacity. At very low temperatures the conductivity begins to fall on decreasing temperature because collisions with the sides of the crystal limit the growth in the mean-free-path length, and the decreasing heat capacity becomes the main determining factor for the temperature dependence.

We need to ask ourselves what types of collisions are responsible for the finite size of the thermal conductivity. It turns out that the main scattering processes are the Umklapp processes. This is because the normal scattering processes ($\mathbf{G} = 0$ in equation (8.3)) will not change the *net* flow of phonons in one direction, so that in the absence of the Umklapp processes the thermal conductivity would be unaffected by the phonon–phonon scattering effects.

Temperature dependence of phonon frequencies

We will consider a simple model, known as the *pseudo-harmonic approximation*, which will give a simple *renormalisation* of phonon frequencies due to the anharmonic terms.[3] The result we will find is that the phonon frequencies

[3] This model is described in more detail in Blinc and Zeks (1974).

vary approximately linearly with temperature, usually *increasing* on increasing temperature. This mechanism is not the only one that causes phonon frequencies to change with temperature. The effects of thermal expansion, which are usually more important, generally cause a *decrease* in phonon frequency on increasing temperature, because the average distance between atoms increases leading to a decrease in the strength of the interatomic interactions. The effects of thermal expansion are best expressed in terms of the Grüneisen parameters that were described in Chapter 5. The importance of the model presented here is that it can predict the existence of displacive phase transitions.

The formal expression for the crystal Hamiltonian including only harmonic and quartic terms obtained from equation (8.1) is:

$$\mathcal{H} = \frac{1}{2}\sum_{\mathbf{k},\nu}\dot{Q}(\mathbf{k},\nu)\dot{Q}(-\mathbf{k},\nu) + \frac{1}{2}\sum_{\mathbf{k},\nu}\omega_0^2(\mathbf{k},\nu)Q(\mathbf{k},\nu)Q(-\mathbf{k},\nu)$$
$$+\frac{1}{4!}\sum_{\mathbf{k},\nu}\sum_{\mathbf{k}',\nu'}\sum_{\mathbf{p},\mu}\sum_{\mathbf{p}',\mu'}\left\{V_4\begin{pmatrix}\mathbf{k},\mathbf{k}',\mathbf{p},\mathbf{p}'\\ \nu,\nu',\mu,\mu'\end{pmatrix}Q(\mathbf{k},\nu)Q(\mathbf{k}',\nu')Q(\mathbf{p},\mu)Q(\mathbf{p}',\mu')\right.$$
$$\left.\times\Delta(\mathbf{k}+\mathbf{k}'+\mathbf{p}+\mathbf{p}')\right\} \tag{8.9}$$

For the present purposes we assume that the quartic terms are the only important anharmonic terms.[4] Moreover, we assume that the atoms undergo small oscillations, so that the anharmonic terms are small in comparison with the harmonic term. This second assumption implies that the character of the phonons does not change significantly in the presence of the anharmonic interactions, and that the only effect of the higher order terms, apart from the finite lifetime effect, is a change in frequency.

The main approximation we make is to replace a pair of normal mode coordinates in equation (8.9) by the thermal averages:

$$Q(\mathbf{p},\mu)Q(\mathbf{p}',\mu') \rightarrow \langle Q(\mathbf{p},\mu)Q(\mathbf{p}',\mu')\rangle \tag{8.10}$$

The thermal average of a pair of normal mode coordinates is zero unless:

$$\mathbf{p}' = -\mathbf{p} \ ; \ \mu' = \mu \tag{8.11}$$

which imposes the conservation requirement for the thermal averages:

[4] The equations are easier to handle in this case, although Bruce and Cowley (1973) have shown that the cubic terms are also important as regards the temperature dependence of the phonon frequencies.

$$\langle Q(\mathbf{p},\mu)Q(\mathbf{p}',\mu')\rangle \propto \delta_{\mathbf{p},-\mathbf{p}'}\delta_{\mu,\mu'} \tag{8.12}$$

This approximation effectively removes some of the fluctuations from the theory, which is the main feature of what is known as a *mean-field theory* (discussed later). In addition, we now only allow scattering processes in which two phonons scatter to form two more; we neglect terms where one phonon breaks into three or three phonons merge into one. Furthermore, we now neglect Umklapp terms.

Bearing in mind the changes in the conservation requirements, the approximate Hamiltonian becomes:

$$\mathcal{H}^{a} = \frac{1}{2}\sum_{\mathbf{k},\nu}\dot{Q}(\mathbf{k},\nu)\dot{Q}(-\mathbf{k},\nu) + \frac{1}{2}\sum_{\mathbf{k},\nu}\omega_0^2(\mathbf{k},\nu)Q(\mathbf{k},\nu)Q(-\mathbf{k},\nu)$$
$$+\frac{1}{4}\sum_{\mathbf{k},\nu}\sum_{\mathbf{p},\mu}V_4\begin{pmatrix}\mathbf{k},-\mathbf{k},\mathbf{p},-\mathbf{p}\\ \nu,\nu,\mu,\mu\end{pmatrix}Q(\mathbf{k},\nu)Q(-\mathbf{k},\nu)\langle Q(\mathbf{p},\mu)Q(-\mathbf{p},\mu)\rangle \tag{8.13}$$

It should be noted that we gain a factor of 6 in the quartic term that comes from the summation over all modes. This can be seen as arising from all allowed scattering processes of the type $(\mathbf{k},\mathbf{k}') \rightarrow (\mathbf{p},\mathbf{p}')$. Two phonons of wave vector $\mathbf{k}$ and $\mathbf{k}'$ scatter from each other to give two new phonons of wave vector $\mathbf{p}$ and $\mathbf{p}'$. In our approximation we include only the wave vectors that obey the criterion $\mathbf{k}+\mathbf{k}'+\mathbf{p}+\mathbf{p}'=0$, and we also require that the terms include only the wave vectors $\mathbf{k}$, $-\mathbf{k}$, $\mathbf{p}$, and $-\mathbf{p}$. Thus we only have the six terms indicated in the table:

$\mathbf{k}$	$\mathbf{k}'$	$\mathbf{p}$	$\mathbf{p}'$
$\mathbf{k}$	$-\mathbf{k}$	$\mathbf{p}$	$-\mathbf{p}$
$\mathbf{k}$	$-\mathbf{k}$	$-\mathbf{p}$	$\mathbf{p}$
$\mathbf{k}$	$\mathbf{p}$	$-\mathbf{k}$	$-\mathbf{p}$
$\mathbf{k}$	$\mathbf{p}$	$-\mathbf{p}$	$-\mathbf{k}$
$\mathbf{k}$	$-\mathbf{p}$	$-\mathbf{k}$	$\mathbf{p}$
$\mathbf{k}$	$-\mathbf{p}$	$\mathbf{p}$	$-\mathbf{k}$

We recall the result from Chapter 4 for the thermal amplitude of a normal mode coordinate in the high-temperature limit:

$$\langle Q(\mathbf{p},\mu)Q(-\mathbf{p},\mu)\rangle = k_B T / \omega^2(\mathbf{p},\mu) \tag{8.14}$$

We can substitute this result into the approximate Hamiltonian given by equation (8.13):

$$\mathcal{H}^{\mathrm{a}} = \frac{1}{2}\sum_{\mathbf{k},\nu}\dot{Q}(\mathbf{k},\nu)\dot{Q}(-\mathbf{k},\nu) + \frac{1}{2}\sum_{\mathbf{k},\nu}\omega_0^2(\mathbf{k},\nu)Q(\mathbf{k},\nu)Q(-\mathbf{k},\nu)$$

$$+\frac{k_{\mathrm{B}}T}{4}\sum_{\mathbf{k},\nu}\sum_{\mathbf{p},\mu}V_4\begin{pmatrix}\mathbf{k},-\mathbf{k},\mathbf{p},-\mathbf{p}\\ \nu,\nu,\mu,\mu\end{pmatrix}Q(\mathbf{k},\nu)Q(-\mathbf{k},\nu)/\omega^2(\mathbf{p},\mu) \quad (8.15)$$

This equation is then equivalent to a harmonic equation, if we replace the harmonic frequencies in equation (8.15) with the modified, or *renormalised*, frequencies $\tilde{\omega}(\mathbf{k},\nu)$:

$$\mathcal{H}^{\mathrm{a}} = \frac{1}{2}\sum_{\mathbf{k},\nu}\dot{Q}(\mathbf{k},\nu)\dot{Q}(-\mathbf{k},\nu) + \frac{1}{2}\sum_{\mathbf{k},\nu}\tilde{\omega}^2(\mathbf{k},\nu)Q(\mathbf{k},\nu)Q(-\mathbf{k},\nu) \quad (8.16)$$

where the renormalised frequencies are given by:

$$\tilde{\omega}^2(\mathbf{k},\nu) = \omega_0^2(\mathbf{k},\nu) + \frac{k_{\mathrm{B}}T}{4}\sum_{\mathbf{k},\nu}\sum_{\mathbf{p},\mu}V_4\begin{pmatrix}\mathbf{k},-\mathbf{k},\mathbf{p},-\mathbf{p}\\ \nu,\nu,\mu,\mu\end{pmatrix}/\tilde{\omega}^2(\mathbf{p},\mu) \quad (8.17)$$

This is called the *pseudo-harmonic approximation*, because we have replaced the anharmonic Hamiltonian by an effective harmonic Hamiltonian, and have replaced the harmonic force constants by effective force constants.[5] The new frequencies are called *renormalised* frequencies because their values have been renormalised by the anharmonic interactions. We can in principle solve equation (8.17) for the new frequencies in two ways. We can put the pseudo-harmonic frequencies in the denominator of equation (8.17) and solve a set of equations for the renormalised frequencies self-consistently (complicated!), or we can replace these renormalised frequencies in the denominator by the harmonic frequencies and calculate the approximate shift in each frequency from its harmonic value.

The important point to note is that we have been able to introduce temperature explicitly into the phonon frequencies. If the coupling constants $V_4(...)$ are approximately independent of temperature, the phonon frequencies vary linearly with temperature. The temperature comes into the picture via the thermal population of the phonons that interact with the phonon whose frequency we are calculating, equation (8.14).

[5] Some authors prefer to use the term *quasi-harmonic* instead of *pseudo-harmonic*. The former term, however, has a number of other uses, one of which occurs later in this chapter, and is generally taken to refer to any model in which the harmonic frequencies are modified by some change in the crystal rather than renormalised by anharmonic interactions. We met an example of this in Chapter 5, where thermal expansion was treated by considering the dependence of phonon frequencies on volume. Yet another name for the model described here is the *independent mode approximation*, as used by Bruce and Cowley (1981, pp 124–128).

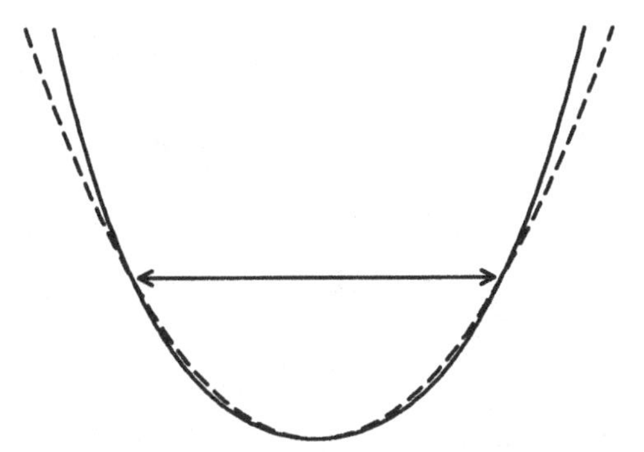

Figure 8.4: Anharmonic potential well for a single atom. The broken curve shows a harmonic term that equals the anharmonic potential at one given energy.

The origin of the positive shift in frequency with temperature is illustrated in Figure 8.4, which shows an anharmonic potential well. As the temperature increases, the anharmonic potential well is narrower than the corresponding harmonic well would be. The atoms are therefore constrained to a tighter amplitude, which gives rise to a higher frequency since the mode amplitude is proportional to ω^{-1}.

Displacive phase transitions and soft modes

The simple picture of a phonon frequency that varies linearly with temperature is very relevant in the standard model of displacive phase transitions. Let us consider a phonon frequency, of wave vector **k**, which has been renormalised by the quartic anharmonic interactions and which can simply be expressed as

$$\tilde{\omega}^2 = \omega_0^2 + \alpha T \tag{8.18}$$

where α will in general be positive.

We recall from our discussion of the calculation of harmonic frequencies that if ω_0^2 is negative the crystal is unstable against the displacements of the corresponding mode eigenvector. We now consider a symmetric high-temperature phase. If it has a harmonic frequency at any wave vector **k** that is imaginary, then the structure is not stable at 0 K, and there is another structure of lower symmetry that has a lower energy at 0 K. The lower energy structure can be viewed as a small modification of the higher-symmetry structure: the modification is caused by the distortion corresponding to the eigenvector of the mode with the imaginary frequency. In other words, the stable structure is equivalent to the symmetric structure with a *frozen-in* normal mode coordinate of wave vector **k** corresponding to the imaginary harmonic frequency

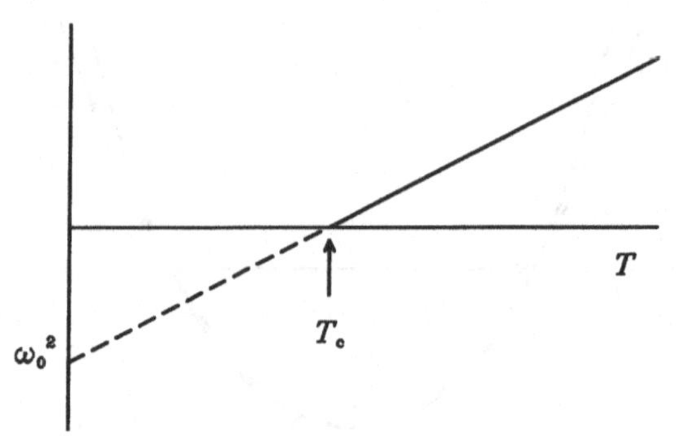

Figure 8.5: Schematic representation of the temperature dependence of the square of the frequency of a soft mode. Below the temperature T_c the frequency is imaginary and hence unstable. The frequency at $T = 0$ K is the harmonic value.

$\omega_0(\mathbf{k}, \nu)$.[6] On warming, the anharmonic contribution to the phonon frequency increases until the renormalised frequency $\tilde{\omega}(\mathbf{k}, \nu)$ becomes zero and then real. At this point the symmetric structure is now stable, and the point at which the renormalised frequency reaches zero in value corresponds to the phase transition between the low-temperature low-symmetry phase and the high-temperature symmetric phase. This gives a transition temperature, T_c, for the phase transition which is related to the fundamental parameters:

$$T_c = -\omega_0^2 / \alpha \tag{8.19}$$

This process is illustrated in Figure 8.5.

Another way of thinking about this mode is that in the high-temperature symmetric phase the mode frequency decreases on cooling until it becomes zero in value. At that point the crystal is unstable against the corresponding distortion and the crystal undergoes a phase transition to a lower-symmetry phase. This mode (in the high-temperature phase) is called a *soft mode*, because it has a low frequency and the crystal is essentially *soft* against the corresponding displacements of the atoms. The frequency is said to soften on cooling towards the transition point. Often the wave vector **k** is a high-symmetry point (a Brillouin zone boundary or the zone centre), but this need not always be so. Also, the transition on cooling occurs as soon as any one point on a phonon branch reaches zero. There is also a soft mode on the low-temperature side of the transition, which increases in frequency on cooling, associated with the instability that occurs on heating.

[6] In practice the calculated frequencies for the high-symmetry structure will occur for a range of wave vectors, and in some cases whole branches may be unstable. Usually though, the instability will occur at the wave vector with the largest imaginary frequency.

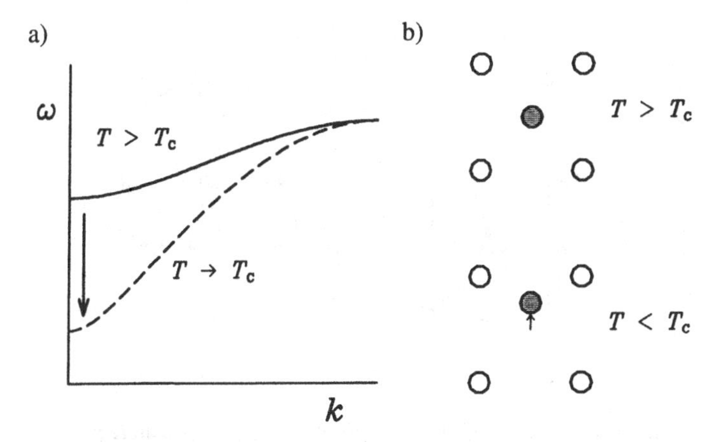

Figure 8.6: Schematic representations of a ferroelectric soft mode behaviour: *a*) behaviour of the phonon dispersion curves; *b*) atomic distortions.

It should be noted that the anharmonic effects that drive a soft mode phase transition need not be particularly strong (Bruce and Cowley 1981, p 91). Rather, the soft mode harmonic frequency has a small imaginary value so that it is particularly sensitive to the effects of the anharmonic interactions. This is because the anharmonic interactions produce a shift in the value of ω^2 rather than a rescaling, and the shifts required to stabilise the soft mode need not be significantly larger than the corresponding changes in the high-frequency modes.

Ferroelectric and zone centre phase transitions

The soft mode theory of phase transitions was originally developed to explain the origins and mechanisms of ferroelectric phase transitions[7] (Cochran 1959a, 1960, 1961). A ferroelectric phase transition involves the loss of a centre of symmetry in the unit cell, and in general this will give rise to a net dipole moment of the unit cell (Lines and Glass 1977). This in turn gives rise to a macroscopic dielectric polarisation of the whole crystal which can be measured fairly easily, just as the alignment of atomic magnetic moments gives rise to an observable magnetisation in a ferromagnetic material. One of the

[7] Cochran (1981) and Dolino (1990) have noted that the soft mode theory had been anticipated many years previously but that the ideas had not been appreciated. The first observation of a soft mode was by Raman and Nedungadi (1940) in quartz, and at the same time Saksena (1940) had predicted the existence of a soft mode in quartz theoretically. The Lyddane–Sachs–Teller (Lyddane et al. 1941) relation was contemporaneous with these studies. The first experimental verifications of the soft mode theory were by Barker and Tinkham (1962) and Cowley (1962).

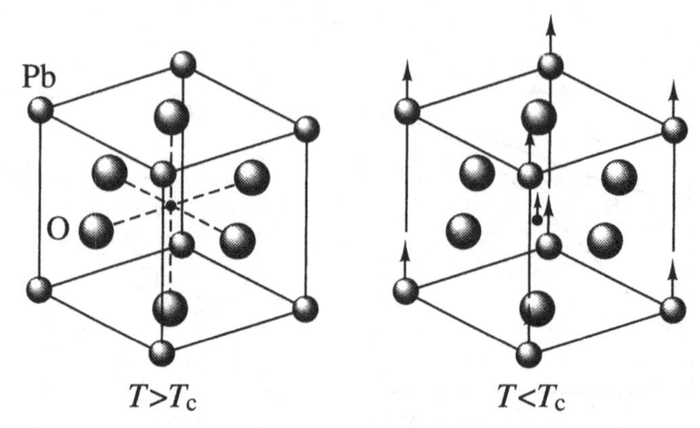

Figure 8.7: The crystal structure of $PbTiO_3$ in the cubic high-temperature phase and the tetragonal low-temperature phase, with the atomic displacements indicated.

characteristic properties of a ferroelectric phase transition is that it is accompanied by a large dielectric constant that increases towards an infinite value at the transition temperature. This effect is in fact predicted by the Lyddane–Sachs–Teller relation (equation (3.29)), which preceded the development of the soft mode theory and which shows that a diverging value of the dielectric constant follows from the softening of a transverse optic phonon frequency with zero wave vector. The atomic motions associated with the soft phonon correspond to the displacements that accompany the ferroelectric phase transition, and these are illustrated schematically in Figure 8.6.

One example of a soft mode ferroelectric phase transition is the perovskite $PbTiO_3$, which undergoes a cubic–tetragonal ferroelectric phase transition at 763 K (Burns and Scott 1970; Shirane et al. 1970). The distortion associated with the transition is shown in Figure 8.7, and primarily involves small atomic displacements along [001]. The transition in $PbTiO_3$ is first order, so the soft mode frequency does not actually reach zero – this is the case for many ferroelectric phase transitions, and is a consequence (discussed in Appendix D) of a coupling to a strain distortion. The temperature dependence of the soft mode has been measured above and below the transition temperature, as shown in Figure 8.8.

A large number of other perovskite crystals also undergo ferroelectric phase transitions. The best known example is $BaTiO_3$, which has a phase transition at 393 K which in some senses is similar to that observed in $PbTiO_3$. However, whereas the soft mode in $PbTiO_3$ is observed to behave as a sharp phonon at all temperatures, the soft mode in $BaTiO_3$ is found to be heavily damped (Yamada et al. 1969; Harada et al. 1971). This is believed to be associated with a degree of disorder in the $BaTiO_3$ structure – rather than the tetragonal phase simply

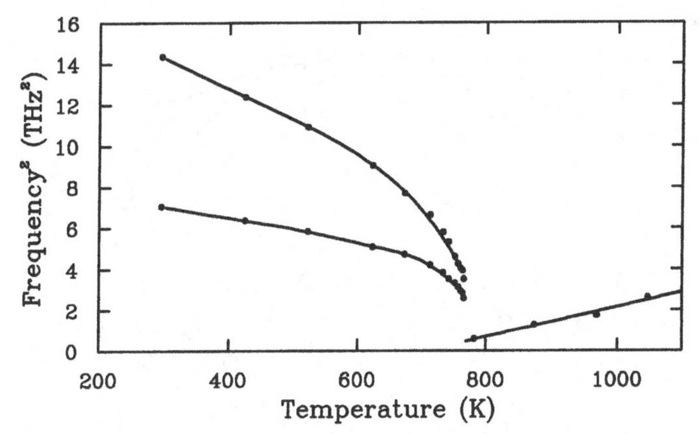

Figure 8.8: Temperature dependence of the soft ferroelectric mode in $PbTiO_3$. The soft mode in the cubic phase is a transverse optic mode, and the square of its frequency varies linearly with temperature. The data in the low-temperature phase were obtained by Raman spectroscopy (Burns and Scott 1970), and the data in the high-temperature phase were obtained by inelastic neutron scattering (Shirane et al. 1970).

consisting of the atoms moving along [001], in $BaTiO_3$ the Ti atoms are located in potential minima lying away from their average position in the cubic phase along the [111] directions. In the tetragonal phase there are then four possible positions for the Ti atoms, and on further cooling there are subsequent structural transitions involving progressive ordering of the Ti atoms, until at low temperatures the structure becomes rhombohedral with the polarisation along [111]. Well-behaved soft modes with zero wave vector are also found in $SrTiO_3$ (Cowley 1962) and $KTaO_3$ (Comès and Shirane 1972; Perry et al. 1989). However, in both of these cases the ferroelectric phase transition appears to try to occur at low temperatures (~32 K in the case of $SrTiO_3$), and at such low temperatures quantum-mechanical effects actually suppress the transition and allow it to occur only at 0 K.

There are also phase transitions that have soft modes with zero wave vector but which are not ferroelectric. One example is quartz, which undergoes a hexagonal–trigonal phase transition at 846 K (Dolino 1990),[8] and although both the high- and low-temperature phases are non-centrosymmetric, there is not an accompanying change in the dielectric polarisation or a divergence of the dielectric constant. The soft mode has recently been measured by neutron scattering (Dolino et al. 1992).

[8] Quartz is actually somewhat more complicated owing to the existence of an incommensurate phase transition at a similar temperature to the transition to the trigonal phase, but the details are only important close to the transition. This aspect will be discussed later in this chapter.

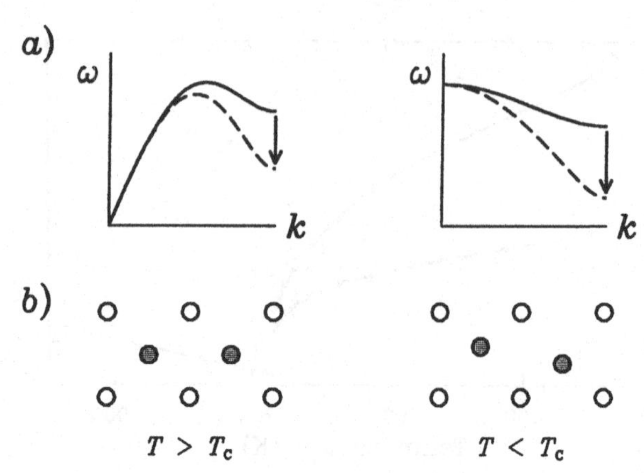

Figure 8.9: Schematic representations of *a*) zone boundary soft acoustic and optic modes, and *b*) atomic displacements showing doubling of the unit cell and the cancelling induced dipole moments.

Zone boundary (antiferroelectric) phase transitions

A number of crystals undergo phase transitions which involve soft modes with wave vectors at Brillouin zone boundaries. In these cases the soft phonons can be either acoustic or optic modes since, as we have seen in Chapter 3, in many cases the distinction between the two types of mode is unclear at zone boundary wave vectors. The different types of soft mode, and the respective atomic displacements, are shown schematically in Figure 8.9. One of the results of a zone boundary soft mode phase transition is that the unit cell of the low-temperature phase is doubled in one or more directions. In some cases neighbouring unit cells of the high-temperature phase develop dipole moments, but as these are in opposite directions the unit cell of the low-temperature has no net moment. By analogy with antiferromagnetism this type of transition is sometimes called an *antiferroelectric* phase transition, although the term is not often used these days.

Undoubtedly the best example of a zone boundary phase transition is the cubic–tetragonal transition in the perovskite $SrTiO_3$, which has a soft mode with wave vector $(\frac{1}{2}, \frac{1}{2}, \frac{1}{2})$ and a transition temperature of 110 K (Cowley et al. 1969; Shirane and Yamada 1969) in addition to the ferroelectric soft mode. The atomic motions that are associated with the soft mode and which freeze into the structure below the transition are shown in Figure 8.10. They mostly consist of rotations of the interconnected TiO_6 octahedra about [001], with neighbouring octahedra in the (001) plane rotating in opposite directions. The rotations in neighbouring (001) planes are in opposite senses, so that the unit cell doubles in each direction (the new low-temperature unit cell is actually an

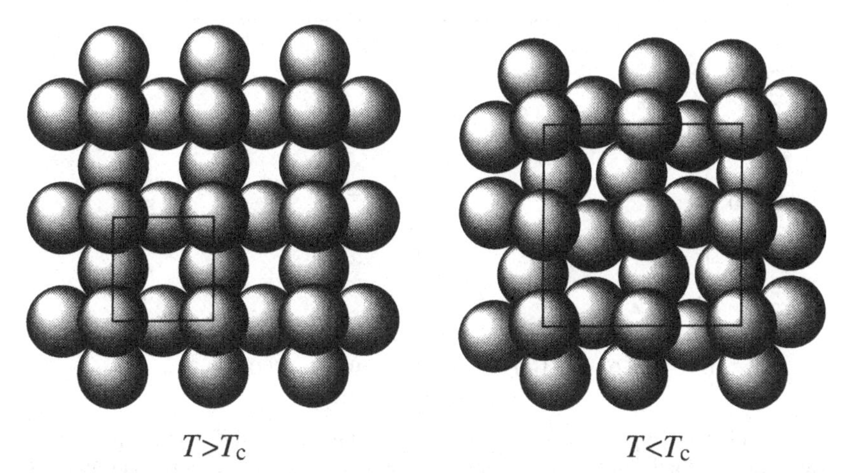

Figure 8.10: The structure change associated with the phase transition in $SrTiO_3$, where we show only the oxygen atoms. The lines outline the unit cells in both phases. In the low-temperature phase the atomic motions associated with the soft mode are identical to the displacement pattern of the low-temperature phase.

unconventional C-centred tetragonal cell, and the cell vectors **a** and **b** of the conventional I-centred cell are at 45° from the corresponding unconventional cell vectors). The temperature dependence of the soft mode on both sides of the transition temperature is shown in Figure 8.11.

Ferroelastic phase transitions

A ferroelastic phase transition is the macroscopic elastic analogue of ferromagnetic or ferroelectric phase transitions, in that the transition involves the creation of a reversible spontaneous shear strain that shows hysteresis behaviour similar to that shown by spontaneous magnetisation or dielectric polarisation (Salje 1990). Ferroelastic phase transitions are accompanied by a combination of elastic constants that falls to zero at the transition temperature, which implies that the gradient of one of the transverse acoustic modes at zero wave vector falls to zero, as discussed in Chapter 7. This is a soft acoustic mode, which may soften across most of the branch or just in the vicinity of zero wave vector. The shear strains generated by transverse acoustic modes are shown in Figures 7.2 and 7.3, and the behaviour of soft acoustic modes is illustrated schematically in Figure 8.12. Experimental data for soft acoustic modes in *sym*-triazine (hexagonal–monoclinic transition at 198 K) and HCN (tetragonal–orthorhombic transition at 170 K) are shown in Figure 8.13.[9] In the case of

[9] In both cases the transition temperatures are 10 K lower in the deuterated forms.

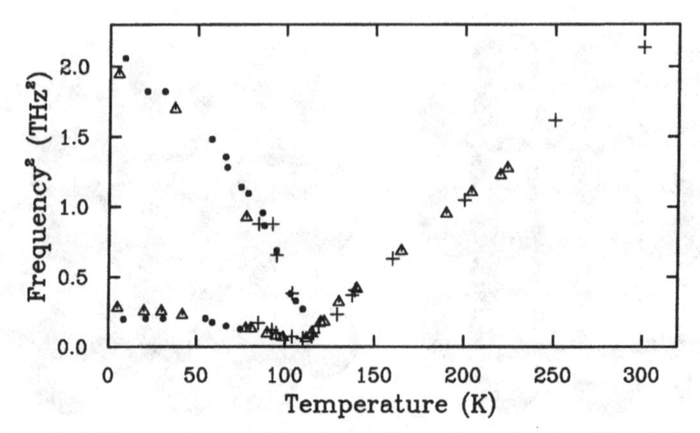

Figure 8.11: The temperature dependence of the soft mode in $SrTiO_3$. The experimental data were obtained from Raman scattering (circles, Fleury et al. 1968) and inelastic neutron scattering (crosses, Cowley et al. 1969; triangles, Shirane and Yamada 1969). The soft mode of the high-temperature phase is triply-degenerate, and splits into a doubly-degenerate and singlet mode in the low-temperature phase.

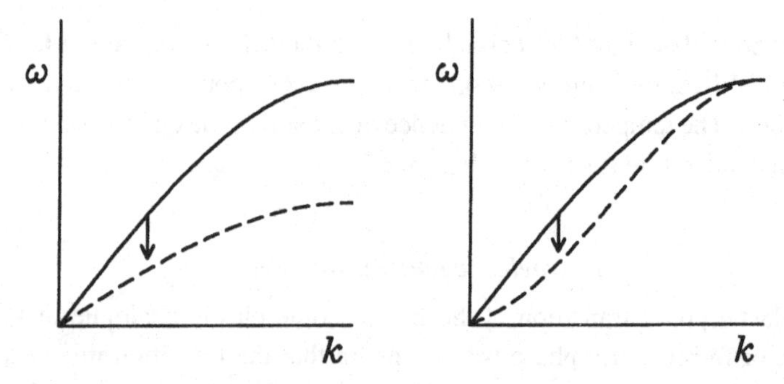

Figure 8.12 Schematic representations of two types of acoustic mode softening, involving the whole or part of the branch.

sym-triazine the softening is confined to wave vectors around $\mathbf{k} = 0$, whereas the whole branch softens in HCN.

Incommensurate phase transitions

In most cases soft modes occur with special values of the wave vector, either at zero or at a zone boundary point. There are a number of cases in which the soft mode falls to zero frequency at a wave vector that is at some point between the zone centre and boundary. The distortions associated with such a soft mode are shown in Figure 8.14. These distortions impose a periodicity on the structure which is unrelated to (incommensurate with) the periodicity of the underlying

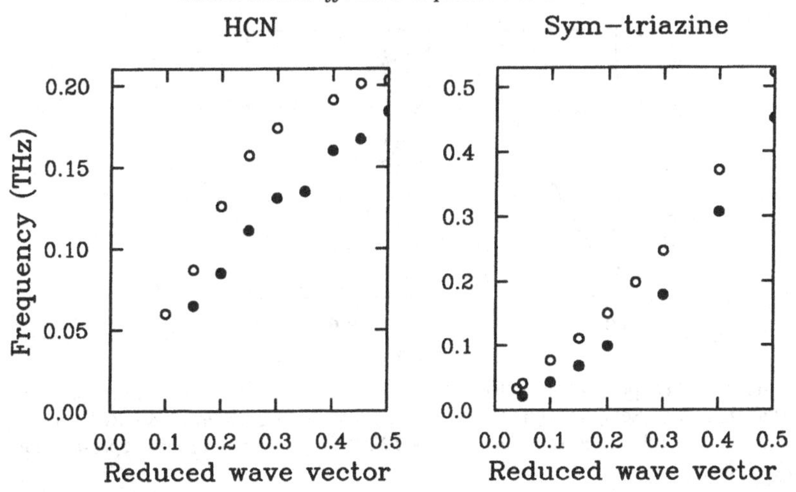

Figure 8.13 Measurements of the soft acoustic branch in deuterated HCN (Mackenzie and Pawley 1979) and deuterated *sym*-triazine (Dove et al. 1983). For HCN the open circles represent data at T_c + 67 K and the closed circles represent data at T_c + 2 K. For *sym*-triazine the open circles represent data at T_c + 107 K and the closed circles represent data at T_c +21 K.

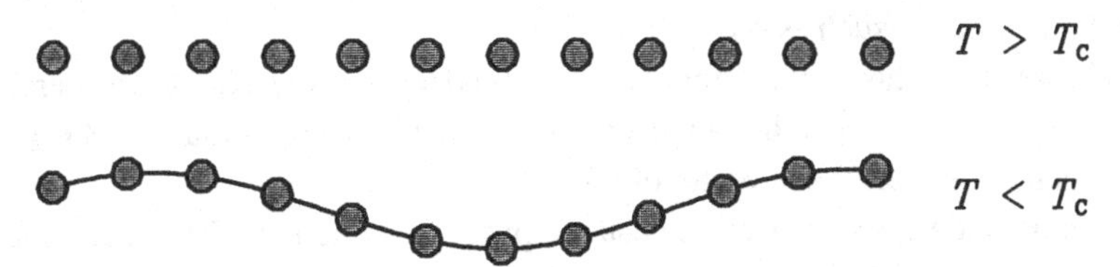

Figure 8.14 Representation of the atomic displacements associated with a soft incommensurate mode.

crystal lattice,[10] and the phase transition is accordingly termed an *incommensurate* phase transition (Blinc and Levanyuk 1986). Another commonly used term for the structure of the incommensurate phase is a *modulated structure*.

There are a number of possible mechanisms for incommensurate phase transitions, one of which is described briefly below in connection with quartz. In most cases an incommensurate phase transition is followed at a lower temperature by a second transition at which the periodicity of the modulation changes so that it is related to the periodicity of the underlying lattice. This transition is known as a *lock-in transition*, since the modulation wave vector has locked in to a value with a periodicity that is commensurate with that of the underlying structure. The lock-in transition can occur just a few degrees below the incom-

[10] Technically the reduced wave vector must not be equal to a fraction of two integers.

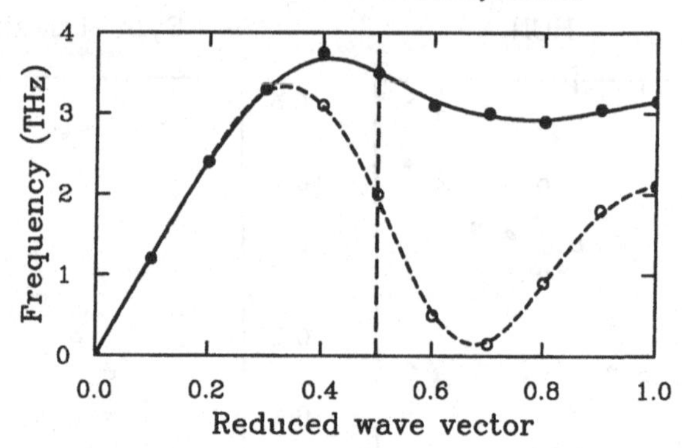

Figure 8.15: Dispersion curve for the branch that softens at the incommensurate phase transition in K_2SeO_4 (Iizumi et al. 1977), shown using an unfolded Brillouin zone. The closed circles represent data at T_c + 122 K and the open circles represent data at $T_c + 2$ K.

mensurate transition, or may occur some hundreds of degrees lower. In general the lock-in wave vector is not very different from the incommensurate wave vector, and it may have a simple value (such as zero, or be at a zone boundary), but in some cases the lock-in wave vector may be some fraction of the reciprocal lattice vector, such as $\frac{1}{3}$ or $\frac{5}{12}$.

Figure 8.15 shows the temperature dependence of the phonon dispersion curve in K_2SeO_4, which has an incommensurate phase transition at 128 K with an incommensurate wave vector of ~0.3**a***.

Quartz undergoes an incommensurate phase transition at 850 K and a lock-in transition at 848 K. For many years the two transitions were not recognised as distinct, and it was believed that the famous α–β phase transition involved a single change. The history of the phase transition in quartz has recently been reviewed by Dolino (1990). The soft optic branch is relatively flat in frequency for wave vectors along [100] (Berge et al. 1986; Bethke et al. 1987; Dolino et al. 1989, 1992; Vallade et al. 1992). As the soft optic mode falls to zero, it interacts with an acoustic mode of the same symmetry for wave vectors along [100] (as anti-crossing effect). However, these two modes have different symmetries at $\mathbf{k} = 0$, so that the interaction strength is zero at $\mathbf{k} = 0$ but increases with increasing wave vector (approximately as k^2). The effect of the interaction is to lower the frequency of the acoustic branch at wave vectors away from the zone centre, and as the temperature is lowered this effect increases until the minimum of the acoustic branch reaches zero.

There are a number of interesting properties of incommensurate phases. One of the main characteristic features is the existence of sharp diffraction peaks that are displaced from the main Bragg peaks by the incommensurate wave vector. These are known as *satellite reflections*, and contain information about the actual structural distortions. There are two new types of lattice vibration that are associated with the incommensurate modulation. The first is associated with fluctuations in the amplitude of the modulation, and the second is associated with fluctuations in the phase of the modulation with respect to the underlying crystal lattice – these are termed *amplitudons* and *phasons* respectively.

Some comments on the experimental aspects of soft modes

The original picture of the mechanism of displacive phase transitions in terms of the softening of a transverse optic phonon has been confirmed by a large number of experimental studies. The early experimental work on the 'model' examples has been summarised by Scott (1974), Shirane (1974) and Axe (1971). The experimental studies on the lattice dynamical aspects of the incommensurate phase transitions pointed to the generality of the idea – a number of the articles in Blinc and Levanyuk (1986) describe measurements of the lattice dynamical aspects of incommensurate phase transitions. Detailed calculations on $SrTiO_3$ by Bruce and Cowley (1973) have confirmed the essential correctness of the basic soft mode model.

The studies on materials such as $BaTiO_3$ showed that not all displacive phase transitions are as simple as the model might suggest – in this case the transition has more of the nature of an order–disorder phase transition, with atoms hopping between sites that are different from the apparent high-symmetry sites. Moreover, in a number of studies the soft phonon is found to dampen on approaching the phase transition simply as a result of the relative enhancement of the anharmonic interactions.

Further experimental studies of $SrTiO_3$ and other materials undergoing second order phase (i.e. continuous, see Appendix D) transitions, however, have shown additional complications with the simple model. The major experimental point, as reviewed by Bruce and Cowley (1981, pp 220–249), is that the soft mode frequency does not reach zero frequency at the transition temperature, but saturates at a finite value. Instead the neutron scattering experiments show the existence of a sharp peak at zero frequency which increases in temperature on approaching the transition temperature from above. Bruce and Cowley (1981, pp 249–316) have reviewed the different possible explanations for this behaviour under three main ideas, all of which are undoubtedly important. The first idea is that the soft mode picture is too simple in its neglect of higher-order anharmonic processes, which should surely become important as the soft mode frequency reaches small values. The second idea is that at temperatures close to but still above the transition temperature the picture of a lattice of atoms vibrating about symmetric positions is inadequate, and instead the crystal contains clusters of the low-temperature structure. The important dynamics in this case concern the motions of the domain walls (called *solitons*) which will occur over much longer time scales than phonon motions. The third idea is that defects are important, particularly close to the transition temperature, and that defects can locally change the energy required for the phase transition to occur. These effects are, however, only important at temperatures close to the transition temperature, and for an understanding of the fundamental origin of the phase transition and the behaviour over a wide range of temperatures we do not need to go beyond the basic ideas of the soft mode model.

Soft modes and the Landau theory of phase transitions

Order parameter and susceptibility

We have noted above that at a soft mode displacive phase transition, the structure of the low-symmetry (low-temperature) phase is equivalent to that of the high-symmetry phase distorted by the static atomic displacements associated with the eigenvector of the soft mode. The distortion pattern is usually more-or-less constant for all temperatures – only the amplitude is significantly temperature dependent. The amplitude of the distortion is called the *order parameter*, and is of central importance for the theory of any phase transition.

The first theoretical approach to any phase transition is *Landau theory*, which is a phenomenological approach that gives useful information concerning a wide range of physical and thermodynamic properties associated with the phase transition (Salje 1990, ch. 13; Blinc and Zeks 1974, ch. 3; Bruce and Cowley 1981, pp 1–74). Landau theory is described in detail in Appendix D. The central idea is that the Gibbs free energy of a crystal,[11] G, can be expressed as a power series expansion in the order parameter, Q, about the free energy of the high-symmetry structure, G_0:

$$G(Q) = G(Q=0) + \frac{1}{2}a(T - T_c)Q^2 + \frac{1}{4}bQ^4 + \cdots \qquad (8.20)$$

From this expansion it is possible to calculate the temperature dependence of the order parameter, as described in Appendix D.

In the high-symmetry phase ($\langle Q \rangle = 0$), Q vibrates as a normal mode. The prefactor, $a(T-T_c)$, is then equivalent to the square of the soft mode frequency as given by equations (8.18) and (8.19). This prefactor is also, in the high-

[11] We follow convention here and describe the phase transition in the language of Landau theory using the Gibbs free energy rather than the Helmholtz free energy.

temperature phase, equivalent to the inverse susceptibility, $\chi^{-1} = \partial^2 G/\partial Q^2$. The soft mode concept as developed in this chapter provides some theoretical justification for the application of Landau theory to displacive phase transitions, and gives some physical interpretation of the phenomenological parameters of equation (8.20).

For a phase transition to occur in a crystal, the potential energy of the crystal (called the lattice energy in previous chapters) must be a double-well function of the order parameter, e.g.,

$$V(Q) = -\frac{1}{2}\kappa_2 Q^2 + \frac{1}{4}\kappa_4 Q^4 + \cdots \tag{8.21}$$

This potential is equivalent to the temperature-independent part of the Landau free energy in equation (8.20). The harmonic prefactor, $-\kappa_2$, is also equivalent to the square of the harmonic component of the soft mode frequency, $\omega_0{}^2$, at $Q = 0$, which we have already pointed out must be negative.

Double-well potentials: a simple model

Considerable theoretical work has been carried out on one simple atomic model that displays a phase transition, which is illustrated in Figure 8.16. The model consists of a lattice of atoms, which are allowed to vibrate along one direction and which interact with their nearest neighbours with harmonic forces (Bruce and Cowley 1981, pp 114–119; Giddy et al. 1989, 1990). Each atom also experiences a force due to a static on-site double-well potential which is assumed to arise from the interaction of the atom with the rest of the crystal. The Hamiltonian for this system is

$$\mathcal{H} = \frac{m}{2}\sum_j \dot{x}_j^2 + \frac{J}{4}\sum_{j,j'}\left(x_j - x_{j'}\right)^2 + \sum_j V\left(x_j\right) \tag{8.22}$$

where

$$V\left(x_j\right) = -\frac{1}{2}ax_j^2 + \frac{1}{4}bx_j^4 \tag{8.23}$$

One key parameter is the depth of the double-well, which is equal to $V_0 = a^2/4b$. The transition is determined by the strength of the coupling parameter J – the higher the value of J the higher the transition temperature. There are two limiting cases of the model. If $V_0 \gg J$, the atoms tend to remain near the minima of the potential wells, i.e. at $\pm x_0$, where $x_0 = (a/b)^{1/2}$, at all

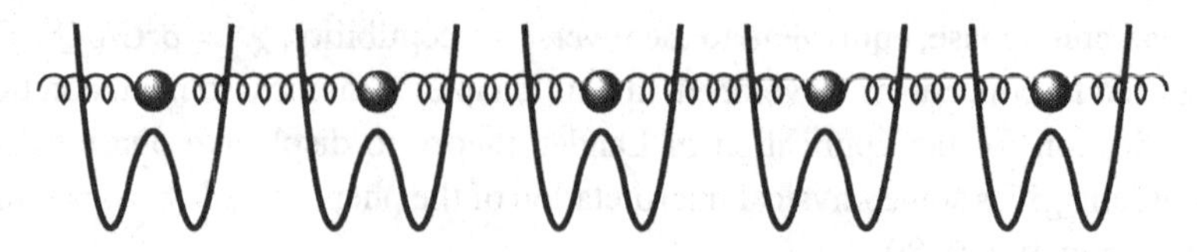

Figure 8.16: Simple atomic model showing nearest-neighbour harmonic interactions and on-site double-well potentials.

temperatures, and the transition involves ordering of the positions of all the atoms to either $+x_0$ or $-x_0$. This transition is of *order–disorder* type, and will not have an associated soft mode. We do not expect Landau theory to be valid for this case.[12] The other limiting case is $V_0 \ll J$. This is a typical soft mode displacive transition, and the theory we have discussed in this chapter is applicable in this case. In practice it turns out that a large number of phase transitions fall into this limiting case, and it is found that Landau theory is appropriate for these transitions over a wide temperature interval.

It is possible to have intermediate cases, where the height of the potential barrier between two minima is similar to $k_B T_c$. It is found in such cases that at temperatures well above the transition temperature the behaviour of the crystal corresponds to the classic soft mode behaviour. On approaching the transition, regions of the crystal fluctuate spontaneously into the low-temperature ordered phase, and the transition takes on characteristics of an order–disorder transition. The soft mode in these cases does not actually reach zero, but instead the lineshape that is measured in an experiment broadens significantly in frequency on cooling. When the width is comparable to the mode frequency, the mode is said to be *overdamped*, since the apparent lifetime of the phonon is no longer than one period of the oscillation. We have previously mentioned overdamped soft modes as observed in $BaTiO_3$.

An exact model of a displacive phase transition

Our aim now is to draw the connection between Landau theory and the soft mode theory of displacive phase transitions – we will find that the lattice dynamical model of the mechanism of a displacive phase transition provides the natural physical basis for the application of Landau theory to these phase transitions. We will use an approach first suggested by Chihara et al. (1973) and subsequently developed by Rae (1982) and reviewed in more general

[12] If mean-field theory is applicable, an order–disorder phase transition may be better described by the Bragg–Williams model (Rao and Rao 1978, pp 184–190), which predicts that the coefficients of an expansion of the free energy in the form of equation (8.27) will be temperature dependent.

terms by Dove et al. (1992a). We start with the standard equation for the free energy of a harmonic crystal:

$$F = k_{\mathrm{B}}T \sum_{\mathbf{k},\nu} \ln\left[2\sinh\left(\frac{\hbar\omega(\mathbf{k},\nu)}{2k_{\mathrm{B}}T}\right)\right] + V \tag{8.24}$$

where V is the potential energy of the crystal, which is added to the normal phonon free energy.

We now use the quasi-harmonic approximation. It is assumed that the phonon free energy is a function of the order parameter Q through the dependence of the harmonic phonon frequencies on Q. We retain only the Q-dependent part of V as given in equation (8.21). The essence of the quasi-harmonic approximation is that the harmonic model is used with new frequencies.

To make things simple, we will assume an Einstein model. For N_{A} atoms, we can rewrite equation (8.24) in the quasi-harmonic approximation as

$$F(Q,T) = 3RT\ln\left[2\sinh\left(\frac{\hbar\omega(Q)}{2k_{\mathrm{B}}T}\right)\right] + V(Q) = F_{\mathrm{ph}}(Q,T) + V(Q) \tag{8.25}$$

We also assume the existence of a double-well potential for $V(Q)$ of the form of equation (8.21). In the high-temperature limit, we can develop the phonon part of $F(Q,T)$, $F_{\mathrm{ph}}(Q,T)$, as an expansion about $F_{\mathrm{ph}}(Q=0)$:

$$F_{\mathrm{ph}}(Q) = F_{\mathrm{ph}}(Q=0) + Q\left(\frac{\partial F_{\mathrm{ph}}}{\partial Q}\right)_{Q=0} + \frac{1}{2}Q^2\left(\frac{\partial^2 F_{\mathrm{ph}}}{\partial Q^2}\right)_{Q=0} \tag{8.26}$$

We can readily obtain the differentials, noting the simplifications to the expressions in the standard high-temperature limit:

$$\frac{\partial F_{\mathrm{ph}}}{\partial Q} = \frac{3}{2}N_{\mathrm{A}}\coth\left(\frac{\hbar\omega}{2k_{\mathrm{B}}T}\right)\hbar\frac{\partial\omega}{\partial Q} \approx \frac{3RT}{\omega}\frac{\partial\omega}{\partial Q} \tag{8.27}$$

$$\begin{aligned}\frac{\partial^2 F_{\mathrm{ph}}}{\partial Q^2} &= \frac{3}{2}N_{\mathrm{A}}\coth\left(\frac{\hbar\omega}{2k_{\mathrm{B}}T}\right)\hbar\frac{\partial^2\omega}{\partial Q^2} - \frac{3}{4}N_{\mathrm{A}}\frac{\hbar^2}{k_{\mathrm{B}}T}\operatorname{cosech}^2\left(\frac{\hbar\omega}{2k_{\mathrm{B}}T}\right)\left(\frac{\partial\omega}{\partial Q}\right)^2 \\ &\approx \frac{3RT}{\omega}\left[\frac{\partial^2\omega}{\partial Q^2} - \frac{1}{\omega}\left(\frac{\partial\omega}{\partial Q}\right)^2\right]\end{aligned} \tag{8.28}$$

We now need to calculate the dependence of the Einstein frequency, ω, on

Q. There are two cases to consider. In the first case, the potential energy of the crystal also contains terms of the form:

$$V(Q,Q_0)=\frac{1}{2}\omega_0^2Q_0^2+\frac{1}{2}\alpha Q_0^2Q^2 \tag{8.29}$$

where ω_0 is the harmonic frequency of the Einstein mode when $Q = 0$, and Q_0 is the normal mode coordinate for the Einstein mode. The potential contains a quartic anharmonic interaction between Q_0 and the normal mode coordinate of the soft mode, Q, which becomes the order parameter in the low-symmetry phase. In the low-symmetry phase, when the average value of Q is non-zero, equation (8.29) can be written as

$$V(Q,Q_0)=\frac{1}{2}\left(\omega_0^2+\alpha Q^2\right)Q_0^2 \tag{8.30}$$

The quasi-harmonic frequency for the Einstein mode is therefore equal to

$$\omega^2=\omega_0^2+\alpha Q^2 \tag{8.31}$$

In this case, therefore, the differentials we require are

$$\frac{\partial\omega}{\partial Q}=\frac{\alpha Q}{\omega}\ ,\ \left(\frac{\partial\omega}{\partial Q}\right)_{Q=0}=0 \tag{8.32}$$

$$\frac{\partial^2\omega}{\partial Q^2}=\frac{\alpha}{\omega}-\frac{\alpha Q}{\omega^2}\frac{\partial\omega}{\partial Q}\ ,\ \left(\frac{\partial^2\omega}{\partial Q^2}\right)_{Q=0}=\frac{\alpha}{\omega_0} \tag{8.33}$$

Substituting equations (8.32) and (8.33) into equation (8.25) gives the final result for the free energy of the crystal in the high-temperature limit as

$$F(Q,T)=\frac{3\alpha RT}{2\omega_0^2}Q^2-\frac{1}{2}\kappa_2Q^2+\frac{1}{4}\kappa_4Q^4 \tag{8.34}$$

This result is equivalent to the Landau free energy of equation (8.20), provided that α is positive. It should be noted that the temperature-dependent term is identical to that obtained from the pseudo-harmonic approximation, equation (8.17) with different notation, which follows directly from the use of the same quartic anharmonic interaction in both cases.

The effect of the order parameter on the phonon frequencies, as given by equation (8.31), is often easily measurable using spectroscopic techniques. Experimental data for the phase transition in As_2O_5 are shown in Figure 8.17. These measurements provide a convenient method for measuring the order

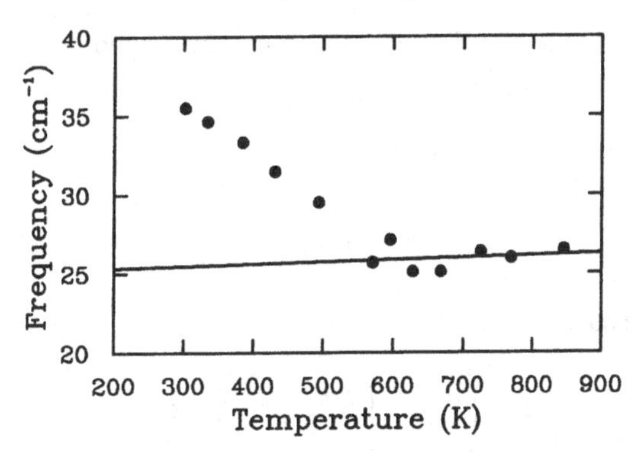

Figure 8.17: The effect of the order parameter on the temperature dependence of a hard mode frequency in As_2O_5, as measured by Raman spectroscopy (Bismayer et al. 1986). The line indicates the extrapolation of the high-temperature data, and highlights the frequency change in the low-temperature phase.

parameter, provided that ω_0, which will have some intrinsic temperature dependence of its own, can be accurately extrapolated from high-temperature measurements. These modes are called *hard modes*, since they do not soften at the transition temperature, and the study of them is called *hard mode spectroscopy* (Bismayer 1988; Salje 1992). Certainly it is often much easier to obtain the temperature dependence of the order parameter from spectroscopic measurements of hard mode frequencies than it is to obtain the same information from crystal structure refinements or other techniques, and the accuracy is often superior also.

The second case we consider is where the Einstein modes correspond to a set of degenerate pairs in the high-symmetry phase, which do not remain degenerate in the low-symmetry phase. The relevant potential is of the form:

$$V(Q, Q_1, Q_2) = \frac{1}{2}\omega_0^2\left(Q_1^2 + Q_2^2\right) + \frac{1}{2}\gamma Q\left(Q_1^2 - Q_2^2\right) \tag{8.35}$$

where we neglect higher-order terms. Q_1 and Q_2 are the normal mode coordinates of the degenerate pair, and ω_0 is the harmonic frequency of the degenerate modes when $Q = 0$. The relevant anharmonic interaction with Q in this case is a cubic term, in which Q_1 and Q_2 interact with Q with opposite sign. The reason for this sign difference is that in the high-symmetry phase, where $\langle Q_1^2 \rangle = \langle Q_2^2 \rangle$, there must be no resultant term that is linear in Q.

In the low-symmetry phase, where $\langle Q \rangle$ has a non-zero value, we can rewrite equation (8.35) as

$$V(Q, Q_1, Q_2) = \frac{1}{2}\omega_1^2 Q_1^2 + \frac{1}{2}\omega_2^2 Q_2^2 \tag{8.36}$$

where ω_1 and ω_2 are the two Einstein frequencies corresponding to the normal mode coordinates Q_1 and Q_2 respectively:

$$\omega_1^2 = \omega_0^2 + \gamma Q \tag{8.37a}$$

$$\omega_2^2 = \omega_0^2 - \gamma Q \tag{8.37b}$$

In this case we have:

$$\left(\frac{\partial \omega_1}{\partial Q}\right)_{Q=0} = -\left(\frac{\partial \omega_2}{\partial Q}\right)_{Q=0} = \frac{\gamma}{2\omega_0} \tag{8.38}$$

$$\left(\frac{\partial^2 \omega_1}{\partial Q^2}\right)_{Q=0} = -\left(\frac{\partial^2 \omega_2}{\partial Q^2}\right)_{Q=0} = \frac{\gamma^2}{4\omega_0^3} \tag{8.39}$$

Substituting equations (8.37) and (8.39) into equation (8.25), recalling that half the modes are for ω_1 and half for ω_2, gives the result:

$$F(Q,T) = -\frac{3\gamma^2 RT}{8\omega_0^4}Q^2 - \frac{1}{2}\kappa_2 Q^2 + \frac{1}{4}\kappa_4 Q^4 \tag{8.40}$$

This is different from the expression for the Landau free energy in that the sign of the temperature-dependent term is necessarily always negative, and hence there is no instability on increasing temperature. The splitting of degenerate modes therefore hinders rather than drives the phase transition.

Although this case does not provide a mechanism for a phase transition, it is nevertheless useful for the measurement of the order parameter. The relationships (8.37a) and (8.37b) give the splitting of the degenerate pair below the transition temperature:

$$\begin{aligned} \omega_1^2 - \omega_2^2 &= 2\gamma Q \\ \Rightarrow \omega_1 - \omega_2 &\approx \gamma Q / \omega_0 \end{aligned} \tag{8.41}$$

This splitting is also easily measured using spectroscopic techniques, and experimental data are shown in Figure 8.18. Measurements of the order parameter from splittings of degenerate modes are usually quite accurate, and unlike the previous case they do not require the extrapolation of ω_0 below the transition temperature.

The analysis of the exact model so far has used the high-temperature approximation in order to show that Landau theory for displacive phase transi-

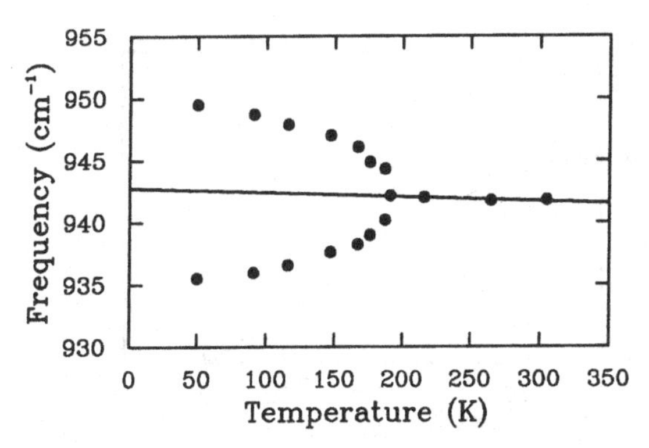

Figure 8.18: The splitting of a degenerate mode below the transition temperature in *sym*-triazine (Daunt et al. 1975). The line is the extrapolation of the high-temperature data.

tions arises quite naturally from exact theories. However, in practice there is no need to make this approximation, and instead we can use equation (8.25) exactly, with appropriate values for the Einstein frequencies. By this means the principal failure of Landau theory, namely the failure to reproduce the correct thermodynamic limiting behaviour that $\partial Q/\partial T = 0$ at $T = 0$ K, is avoided.

Let us put in some numbers. If we take the parameter values, $\omega_0/2\pi = 3.61$ THz, $\kappa_2 = 1.0$ J mol^{-1}, $\kappa_4 = 0.8224$ J mol^{-1}, $\alpha = 4.14 \times 10^{22}$ s^{-2}, $\gamma = 0$, we have the following values for the coefficients in the Landau free energy:

$$a = \frac{3\alpha R}{\omega_0^2} = 2.0 \times 10^{-3}\,\mathrm{J\ mol^{-1}} \tag{8.42a}$$

$$T_c = \frac{\kappa_2 \omega_0^2}{3\alpha R} = 500\ \mathrm{K} \tag{8.42b}$$

$$b = \kappa_4 = 1.0\ \ \mathrm{J\ mol^{-1}} \tag{8.42c}$$

The temperature dependence of the order parameter calculated from the Landau expansion is compared with the exact calculation in Figure 8.19. It can be seen that there is good agreement between the two curves at most temperatures, but below about 100 K the order parameter in the exact calculation flattens off (the effect known as *order parameter saturation*), whereas the Landau model predicts that the order parameter will continue to rise on cooling (Salje et al. 1991). The thermodynamic condition that $\partial Q/\partial T = 0$ at $T = 0$ K is not built in to the Landau model since it is a high-temperature approximation to the exact quantum-mechanical free energy. An example of an actual

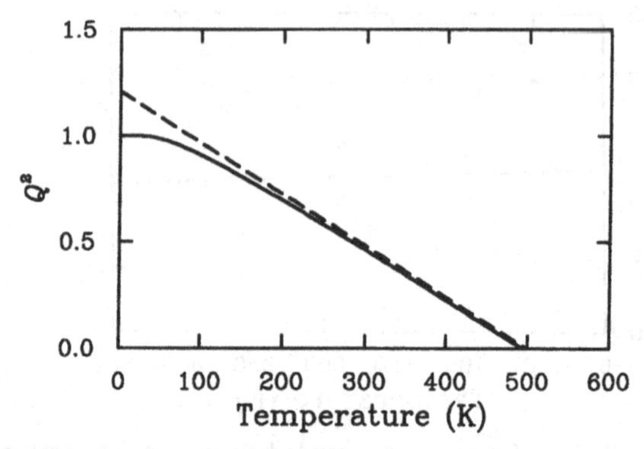

Figure 8.19: Calculated temperature dependence of the order parameter, Q, in the model described in the text. The broken curve is the result from the Landau approximation.

application of the exact model is given by the analysis of the ferroelastic phase transitions in the molecular crystals *sym*-triazine (Rae 1982) and NaN_3 (Aghdaee and Rae 1983). In these cases, each mode used in the analysis was degenerate in the high-temperature phase, and it was found that both cubic and quartic anharmonic interactions were required to account for the observed behaviour. From measurements of the temperature dependence of the order parameter and the phonon frequencies at $\mathbf{k} = 0$ it was possible to obtain values for ω_0, α and γ for each Einstein mode considered. Thus it was possible to construct the potentials given by equations (8.29) and (8.37), as merged into a single potential, exactly. From the final equation for the free energy it was possible to go back and calculate the temperature dependence of the order parameter in both cases, which turned out to agree quite closely with the experimental data. The model also predicted the transition temperatures to within 1 K of the actual transition temperature.

Validity of Landau theory

Landau theory was originally developed to explain behaviour close to T_c, where a series expansion of the free energy might be expected to be valid. As we have shown in this chapter, the lattice dynamics interpretation of Landau theory indicates that when higher-order anharmonic interactions are negligible compared to the quartic terms, Landau theory should be expected to work over a wide temperature range. Moreover, since the effects of the anharmonic interactions increase with temperature, it would be expected that this condition will be met at temperatures well below the transition temperature.

Where Landau theory often does fail in the general case – particularly for order–disorder transitions with short-range interactions, such as magnetic transitions (Fisher 1983), but also for some displacive phase transitions (Bruce and Cowley 1981, pp 112–217) – is in the region for which it was supposed to be most applicable, namely close to the transition temperature. In this region the effects of fluctuations that are neglected by Landau theory become important. The neglect of these fluctuations is equivalent to the assumption that each atom in the crystal sees the same local environment (a *mean-field*). Landau theory is thus an example of a *mean-field theory*, which in the general sense is a theory in which each relevant variable is assumed to see an average environment.[13] This mean-field approximation is a very powerful technique, since it makes many difficult problems tractable. However, when fluctuations are important, as in a magnetic phase transition, mean-field theory fails spectacularly. In such cases mean-field theory overestimates the value of T_c and gives a poor representation of the temperature dependence of the order parameter.[14] An alternative description of the condition for the suitability of mean-field theory is that the correlations associated with the new ordering are intrinsically long-range, so that the continuity of the local environment is forced upon the system. For example, in the simple model described by equations (8.22) and (8.23), if $J \gg V_0$, the phonon dispersion curve is very steep compared with the energy of the double well. Thus the amplitudes associated with the minimum point on the dispersion curve (the soft mode) are much larger than those for modes with other wave vectors. In a real space picture this means that the atoms prefer to move in the local double-well potential in a cooperative manner. The fact that the soft regime on the dispersion curve is in a very narrow range of wave vectors is equivalent to saying that the correlations are long-range, and hence Landau theory as a mean-field theory is expected to be applicable (Dove et al. 1992a). The phase transition in $SrTiO_3$ is known to be dominated by fluctuations, and Landau theory is not well obeyed in the vicinity of the transition temperature (Müller and Berlinger 1971; Riste et al. 1971).

[13] In a magnetic system, the mean-field approximation is the assumption that each spin experiences an average field generated by all the other spins in the system. In a site disordered model, as in a binary alloy, the mean-field approximation is the assumption that each atom is surrounded by the same number of neighbours of each atom type. In both these examples the mean-field approximation neglects local configurations.

[14] Mean-field theories universally predict that $Q \propto (T_c - T)^\beta$, with $\beta = \frac{1}{2}$, whereas for phase transitions that are dominated by fluctuations it is generally found that the value of β is nearer $\frac{1}{3}$, e.g. 0.38.

The origin of the anharmonic interactions

Logically the question of the origin of the anharmonic terms in equation (8.1) comes rather earlier in the story than the end, but with a knowledge of the effects of anharmonic interactions we can ask a rather more fundamental question. The real interatomic interactions are obviously anharmonic in nature; for example, both the Coulombic interaction and the short-range Born–Mayer interactions will contribute to the anharmonic terms. However, for the existence of phase transitions we are interested in the more specific question of the origin of the double minimum in the potential energy, as expressed in simplest form by equation (8.21). This is still an open question! In some cases a simple model potential contains a sufficiently correct representation of the anharmonic interactions to be able to calculate properties associated with a phase transition. In this regard recent work on quartz, in which a simple Buckingham potential with Coulombic interactions within the rigid-ion approximation was developed from *ab initio* quantum-mechanical calculations, provides a good example (Tsuneyuki et al. 1988, 1990; Tautz et al. 1991). However, for materials that display ferroelectric phase transitions a shell-model description may be essential. There is an idea that for these materials the interaction between the core and shell of the O^{2-} ion may be anharmonic, and it may be that this interaction provides the important anharmonic interactions. The justification for this idea lies in the fact that the O^{2-} ion is intrinsically unstable, and is stabilised in the solid state only by the crystal fields. Thus a double-well core–shell interaction may reflect this instability. These ideas are as yet untested, and at the present time there is a large amount of effort being spent on attempting to calculate the interatomic forces in simple ferroelectric materials by *ab initio* quantum-mechanical methods.

Summary

1 The harmonic approximation is found to fail in a number of key aspects, particularly as temperature is increased.
2 Anharmonic interactions are essential for a realistic model of thermal conductivity.
3 Anharmonic interactions have two effects on phonon frequencies. The first is an indirect effect: the anharmonic interactions give rise to thermal expansion which causes phonon frequencies to decrease. The second is a direct effect that acts to increase phonon frequencies. This second effect is present even when thermal expansion is absent, and is involved in the mechanism of displacive phase transitions.
4 Displacive phase transitions are associated with a soft mode, which is a

phonon mode of the high-temperature phase that is only stabilised by anharmonic interactions. The soft mode frequency decreases to a zero value on cooling towards the transition temperature.

5 The soft mode model is described in relation to a number of different types of displacive phase transitions.

6 Anharmonic phonon theory and the soft mode model are shown to account for the successful application of Landau theory to displacive phase transitions.

FURTHER READING

Ashcroft and Mermin (1976) ch. 25, 27
Blinc and Levanyuk (1986)
Blinc and Zeks (1974)
Brüesch (1982) ch. 5
Brüesch (1987) ch. 2–4
Cochran (1973) ch. 6
Iqbal and Owens (1984)
Kittel (1976) ch. 5, 13
Lines and Glass (1977)
Rao and Rao (1978)
Salje (1990) ch. 1–6, 10

9

Neutron scattering

This chapter deals with the use of coherent inelastic neutron scattering for the measurement of phonon dispersion curves. The properties of neutrons and the production of neutrons are described. The general formalism of scattering theory that is presented in Appendix E is developed and applied to the measurement of individual normal modes. Two different approaches, using steady state and pulsed neutron sources, are described.

Properties of the neutron as a useful probe

The fact that particles have wave properties means that we can use particle beams to study the microscopic behaviour of matter. One common example is the use of electron diffraction. We can also scatter beams of neutrons from matter, and it turns out that neutrons have very nice properties for this purpose. The energy, E, and momentum, $\mathbf{p}$, of a neutron are related to its wave vector $\mathbf{k}$ (= 2π/*wavelength*) by

$$E = \frac{\hbar^2 k^2}{2m} \;;\; \mathbf{p} = \hbar \mathbf{k} \tag{9.1}$$

where m is the neutron mass. There are a number of useful properties of neutrons with thermal energies.

1 The mass of the neutron (1.675×10^{-27} kg) is of the same order as the mass of a nucleus. Hence the neutrons can scatter elastically (no change in energy) or inelastically (with energy change) from the nuclei.
2 The wavelength of a neutron beam is typically in the range ~1–5 Å. This means that the neutron beam has good wave vector resolution for studies over the length scales at the unit cell level, comparable with that of X-rays (with wavelengths in the range ~1–2 Å).

3 Thermal neutrons will have energies of up to ~100 meV,[1] which are similar to phonon energies. For example, a neutron beam of energy 20.68 meV (3.31×10^{-18} J per neutron) will have a wavelength of 2.0 Å, and a corresponding phonon frequency of 5.0 THz. This energy corresponds to a temperature of 240 K. The fact that the neutron wavelength and energy are so close to typical values for phonons arises from the size of the neutron mass. X-ray frequencies are much higher than phonon frequencies (for Cu K_α radiation the frequency is 1.95×10^6 THz), and it is generally too difficult to obtain sufficient energy resolution of an X-ray beam to be able to use X-rays to measure phonon frequencies.[2] Electromagnetic radiation with frequencies around the visible region can be used to measure phonon frequencies, but in this case the wavelength is so much larger than phonon wavelengths that measurements are restricted to phonon wave vectors close to the centre of the Brillouin zone (Chapter 10).
4 The electrostatic charge on the neutron is zero, which means that the neutrons do not interact with the electrons in matter via electrostatic forces. Thus for materials that do not have any magnetic order (short- or long-range) the neutrons are not coherently scattered by the electrons.
5 Neutrons interact with atomic nuclei via the strong nuclear force. This is a short-range interaction, with an interaction length of the order of 10^{-15} m. This distance is much shorter than typical interatomic distances and the wavelength of thermal neutrons, so the nuclei effectively behave as point particles for scattering of thermal neutrons. This means that there is no variation of the scattering amplitude with scattering angle.
6 The neutron has a magnetic moment, which arises from the internal quark substructure of the neutron. Therefore the neutron is able to interact with the magnetic moments of the atoms inside a crystal (which arise from the electronic structure); we will not consider this any further.
7 Because the interactions are only with the nuclei, for most materials there is only a very low absorption of the neutron beam. This means that it is easy to control the sample environment (temperature or pressure) since the neutrons are able to pass through the walls of sample chambers. It also means that neutrons will be scattered from the bulk of the material being studied. This is unlike the case of X-rays, which are strongly absorbed by most materials, and which for crystals bigger than ~1 mm^3 are scattered primarily from the surface. It proves to be fortunate that there are some

[1] Neutron energies are typically given in units of meV. Given that it is energy changes that are measured in neutron scattering, it is also common for phonon dispersion curves to be presented in units of meV rather than THz.

[2] Sufficiently high resolution can only be obtained using synchrotron radiation.

materials (e.g. boron and cadmium) which are strong neutron absorbers; these can be used to mask materials that necessarily lie in the neutron beam, so that there is no scattering from them.

These properties mean that neutrons provide an ideal probe of lattice vibrations. In fact neutron scattering is the *only* available technique for measuring phonon dispersion curves across the whole Brillouin zone. In this chapter we will be concerned with *coherent* neutron scattering, in which each nucleus of the same atom type in each unit cell contributes an equal amount of scattering power. When there is a significant dependence of the scattering on the nuclear spin state, or when significant fractions of different isotopes of the same atom are present, there is an additional *incoherent* scattering intensity due to the fact that the different nuclei of the same atom type scatter differently. Coherent and incoherent scattering are sufficiently different that they are treated separately. We will consider only coherent scattering here; incoherent scattering is discussed in some detail by Bée (1988).

Sources of thermal neutron beams

Reactor sources

The most common source of neutrons is a nuclear reactor which has been designed for the production of neutron beams rather than power. Neutrons are produced by the fission reactions within the fuel elements inside the reactor, and are immediately slowed down to lower energies by collisions with atoms in a moderator that surrounds the fuel. In a power reactor the neutrons are used to initiate further fission reactions, whereas in a research reactor the neutrons are scattered from within the reactor down tubes that leave the reactor through beam holes. The beam of neutrons that emerges from the beam hole has a spectrum of energies that is determined by the temperature of the moderator within the reactor, which is typically around room temperature. Higher- or lower-energy beams can be produced by scattering the neutrons from a vessel containing hot (usually heated graphite at ~ 2000 K) or cold (liquid hydrogen at ~ 30 K) material.

The neutron scattering instruments are sited at the positions of the beam holes. Careful shielding is required in order to absorb dangerous high-energy radiation, such as fast neutrons or gamma rays, which apart from the risk to health also give too high a background level in an experimental measurement.

Spallation sources

Spallation neutron sources have recently become available as competitive alternatives to the use of reactor sources. The principle of a spallation source is that the beam of neutrons is produced when a beam of high-energy protons strikes a heavy metal target (e.g. ^{238}U). The proton beam is produced as a pulse in a moderate-sized accelerator. Each proton pulse creates a single pulse of neutrons with a broad energy spectrum, which can be cooled to thermal energies using an appropriate moderator. The pulse nature of the beam, which contains neutrons of a wide range of energies, and the fact that neutrons of different energies travel at different speeds can be exploited by building instruments that measure the time taken by the neutrons in a single pulse to leave the target, pass through the experiment, and reach the detector. The neutron flux from a spallation source is considerably weaker than from a reactor source, but this is offset by the ability to use all the neutrons within a single pulse with time-of-flight techniques.

Control of the neutron beam

When high beam flux is the principal criterion, the neutron scattering instrument will be positioned at the face of the reactor. However, it is often preferable to be able to locate the instrument some distance away from the reactor, either in order to decrease the background radiation or in order to be able to place more instruments on the beam lines than could be accommodated bunched around the reactor face. It is possible to guide neutron beams along tubes (called *guide tubes*) without a major loss in intensity by exploiting the fact that neutrons can reflect from surfaces if the glancing angle is small enough.

Interactions of neutrons with atomic nuclei

Neutrons are scattered from any object with a change in wave vector **Q** (called the *scattering vector* or *wave vector transfer*), and a change in energy $\hbar\omega$ (called the *energy transfer*) which has a value of zero for *elastic scattering* and a non-zero value (positive or negative) for *inelastic scattering*. The intensity of scattering of any radiation from any single object will be a function of the scattering vector **Q**. This **Q**-dependence will be determined by the Fourier transform of the interaction potential, which is often related to the density of the object that scatters the radiation. Since X-rays scatter from the electron density of an atom, the intensity will be strongly dependent on the magnitude of **Q**, as the X-ray wavelength is of the same order of size as the radius of the atom. However, in the case of neutron scattering, the interaction potential occurs over a length scale of $\sim 10^{-15}$ m, which is considerably shorter than the wavelength of the neutron beam and of the length scales that are probed by neutron scattering. Thus the appropriate interaction potential is approximately a delta function, which has a Fourier transform equal to a constant value. Hence the intensity of neutron scattering is independent of the scattering vector.

The strength of the interaction between the neutron and an atomic nucleus is conventionally expressed as a cross-sectional area, $\sigma = 4\pi b^2$, where b is the *scattering length*. The scattering length therefore enters the theory of neutron scattering developed in Appendix E as the quantity that determines the amplitude of the scattered neutron beam.

Whereas the corresponding quantities in the X-ray case (called the *atomic form factors* or *scattering factors*) vary monotonically with atomic number, the scattering lengths in the case of neutron scattering vary erratically from one atom type to another. Moreover, the scattering lengths for most nuclei are not as dissimilar as they are for X-rays. Some representative values of neutron scattering lengths are given below (in units of fm = 10^{-15} m, data from Sears (1986, 1992), as reproduced in Bée (1988, pp 17–27)).

Atom	*b*	*Atom*	*b*	*Atom*	*b*
H	−3.74	O	5.81	P	5.13
D	6.67	Mg	5.38	S	2.85
C	6.65	Al	3.45	Mn	−3.73
N	9.36	Si	4.15	Fe	9.54

Note that the scattering lengths can be either positive or negative, since nuclei can scatter neutrons with or without a change in phase (π) of the neutron wave function. An interesting case is given by the two isotopes of hydrogen, which scatter with opposite sign. The theory of the interaction between neutrons and nuclei is given by Sears (1986).

The neutron scattering function

Summary of main results

The main general results of scattering theory are derived within the classical approximation in Appendix E. An incoming beam of neutrons with wave vector $\mathbf{k}_i$ and energy E_i will be scattered with final wave vector $\mathbf{k}_f$ and final energy E_f. We show in Appendix E that the intensity of a scattered beam of neutrons with scattering vector $\mathbf{Q} = \mathbf{k}_i - \mathbf{k}_f$ and energy transfer $\hbar\omega = (E_i - E_f)$ is proportional to the *scattering function* $S(\mathbf{Q}, \omega)$, which is defined as

$$S(\mathbf{Q},\omega) = \int F(\mathbf{Q},t)\exp(-i\omega t)\mathrm{d}t \tag{9.2}$$

where the *intermediate scattering function*, $F(\mathbf{Q}, t)$, is given as

$$F(\mathbf{Q},t) = \langle \rho(\mathbf{Q},t)\rho(-\mathbf{Q},0)\rangle \tag{9.3}$$

and the *density function* $\rho(\mathbf{Q}, t)$ (sometimes called the *density operator*) is defined as the Fourier transform of the instantaneous nuclear density weighted by the scattering length:

$$\rho(\mathbf{Q},t) = \sum_j b_j \exp\left(i\mathbf{Q}\cdot\mathbf{r}_j(t)\right) \tag{9.4}$$

The instantaneous position of the j-th nucleus, $\mathbf{r}_j(t)$, can be written as

$$\mathbf{r}_j(t) = \mathbf{R}_j + \mathbf{u}_j(t) \tag{9.5}$$

where $\mathbf{R}_j$ is the equilibrium (or mean) position of the nucleus and $\mathbf{u}_j(t)$ is the instantaneous displacement.

The full equation for the scattering function follows from these equations:

$$S(\mathbf{Q},\omega) = \sum_{i,j}\left\{b_i b_j \exp\left(i\mathbf{Q}\cdot\left[\mathbf{R}_i - \mathbf{R}_j\right]\right)\right.$$
$$\left.\times\int\left\langle \exp\left(i\mathbf{Q}\cdot\left[\mathbf{u}_i(t) - \mathbf{u}_j(0)\right]\right)\right\rangle \exp(-i\omega t)\mathrm{d}t\right\} \tag{9.6}$$

This expression includes diffraction by the crystal (no phonons involved in the scattering process) and scattering involving one, two or more phonons.

Expansion of the neutron scattering function for harmonic crystals

All of the dynamic information in equation (9.6) is contained in the time correlation function $\mathcal{G}_{ij}(\mathbf{Q}, t)$:

$$\mathcal{G}_{ij}(\mathbf{Q},t) = \left\langle \exp\left(i\mathbf{Q}\cdot\left[\mathbf{u}_i(t) - \mathbf{u}_j(0)\right]\right)\right\rangle \tag{9.7}$$

We can make more progress if we examine this function in more detail. It is a standard result for a harmonic oscillator (whether classical or quantum) that thermal averages involving two variables (or operators in the quantum case) A and B can be expressed as (Ashcroft and Mermin 1976, p 792; Squires 1978, pp 28–30):

$$\left\langle \exp(A)\exp(B)\right\rangle = \exp\left(\left\langle (A+B)^2\right\rangle / 2\right) \tag{9.8}$$

Although crystals are only approximately harmonic, the behaviour of an anharmonic crystal is often close to that of a modified harmonic crystal (Chapter 8). In any case, the result (9.8) relies on the distribution functions of the two variables being Gaussian, which will generally be approximately true. The correlation function in equation (9.7) can then be re-expressed using equation (9.8):

$$\left\langle \exp\left(i\mathbf{Q}\cdot\left[\mathbf{u}_i(t)-\mathbf{u}_j(0)\right]\right)\right\rangle$$

$$= \exp\left(-\frac{1}{2}\left\langle[\mathbf{Q}\cdot\mathbf{u}_i(t)]^2\right\rangle - \frac{1}{2}\left\langle[\mathbf{Q}\cdot\mathbf{u}_j(0)]^2\right\rangle + \left\langle[\mathbf{Q}\cdot\mathbf{u}_i(t)][\mathbf{Q}\cdot\mathbf{u}_j(0)]\right\rangle\right) \tag{9.9}$$

The first two terms in the right hand exponent are the time-independent Debye–Waller factors:

$$\left\langle[\mathbf{Q}\cdot\mathbf{u}_i(t)]^2\right\rangle = \left\langle[\mathbf{Q}\cdot\mathbf{u}_i(0)]^2\right\rangle = 2W_i \ ; \ \left\langle[\mathbf{Q}\cdot\mathbf{u}_j(0)]^2\right\rangle = 2W_j \tag{9.10}$$

Therefore the correlation function $\mathcal{G}_{ij}(\mathbf{Q}, t)$ is given as

$$\mathcal{G}_{ij}(\mathbf{Q},t) = \exp\left(-\left[W_i + W_j\right]\right)\exp\left\langle[\mathbf{Q}\cdot\mathbf{u}_i(t)][\mathbf{Q}\cdot\mathbf{u}_j(0)]\right\rangle \tag{9.11}$$

We can now expand the exponential term of equation (9.11) as a power series:

$$\exp\left\langle[\mathbf{Q}\cdot\mathbf{u}_i(t)][\mathbf{Q}\cdot\mathbf{u}_j(0)]\right\rangle = \sum_{m=0}^{\infty}\frac{1}{m!}\left\langle[\mathbf{Q}\cdot\mathbf{u}_i(t)][\mathbf{Q}\cdot\mathbf{u}_j(0)]\right\rangle^m \tag{9.12}$$

which leads to the results:

$$S(\mathbf{Q},\omega) = \sum_{m=0}^{\infty} S_m(\mathbf{Q},\omega) \tag{9.13}$$

$$S_m(\mathbf{Q},\omega) = \frac{1}{m!}\sum_{i,j}\left\{b_i b_j \exp\left(i\mathbf{Q}\cdot\left[\mathbf{R}_i - \mathbf{R}_j\right]\right)\exp\left(-\left[W_i + W_j\right]\right)\right.$$
$$\left.\times\int\left\langle[\mathbf{Q}\cdot\mathbf{u}_i(t)][\mathbf{Q}\cdot\mathbf{u}_j(0)]\right\rangle^m \exp(-i\omega t)\mathrm{d}t\right\} \tag{9.14}$$

The terms for different values of m each have a different but significant interpretation. We will see below that we do not need to worry about the convergence of this series, as the important information is contained in the lowest two terms, $m = 0$ and $m = 1$.

The case m = 0: Bragg scattering

For the case $m = 0$, the correlation function given by equation (9.12) is simply a constant of value unity and is thus independent of t. Therefore the time Fourier transform of equation (9.12) will give a delta function at $\omega = 0$, so that the $m = 0$ contribution to the scattering function will be:

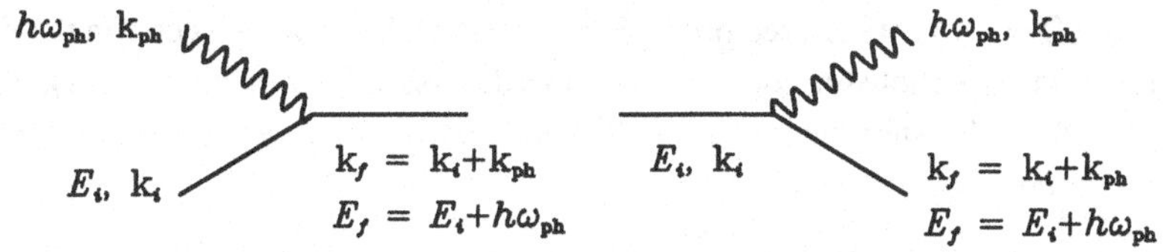

$\hbar\omega_{ph}$, k_{ph} $\qquad$ $\hbar\omega_{ph}$, k_{ph}

E_i, k_i $\quad k_f = k_i + k_{ph}$, $E_f = E_i + \hbar\omega_{ph}$ $\qquad$ E_i, k_i $\quad k_f = k_i + k_{ph}$, $E_f = E_i + \hbar\omega_{ph}$

Figure 9.1: Scattering of a neutron by one phonon, showing scattering involving absorption of a phonon (left) and scattering involving creation of a phonon (right).

$$S_0(\mathbf{Q},\omega) = \sum_{i,j} b_i b_j \exp\left(i\mathbf{Q}\cdot\left[\mathbf{R}_i - \mathbf{R}_j\right]\right)\exp\left(-\left[W_i + W_j\right]\right)\delta(\omega) \quad (9.15)$$

This contribution is purely elastic and involves scattering from no phonons (the change in frequency is zero). In an experiment we would measure the integral over energies, so that equation (9.15) reduces to

$$S_0(\mathbf{Q}) = \left|\sum_j b_j \exp\left(i\mathbf{Q}\cdot\mathbf{R}_j\right)\exp\left(-W_j\right)\right|^2 \quad (9.16)$$

Thus the elastic scattering is equivalent to Bragg scattering, and is only non-zero when $\mathbf{Q}$ is equal to a reciprocal lattice vector. It is important to note that the definition of elastic scattering is not that $S(\mathbf{Q}, \omega=0) \neq 0$ but that $S(\mathbf{Q}, \omega) \propto \delta(\omega)$.

The case m = 1: single-phonon scattering

The term for $m = 1$ involves the interaction between the neutron and one phonon; the neutron either absorbs a phonon and is scattered with a gain in energy, or else the neutron creates a phonon and is scattered with a loss in energy. These two processes are illustrated in Figure 9.1. The $m = 1$ term of equation (9.14) is equal to:

$$S_1(\mathbf{Q},\omega) = \sum_{i,j}\Big\{ b_i b_j \exp\left(i\mathbf{Q}\cdot\left[\mathbf{R}_i - \mathbf{R}_j\right]\right)\exp\left(-\left[W_i + W_j\right]\right)$$
$$\times \int \left\langle\left[\mathbf{Q}\cdot\mathbf{u}_i(t)\right]\left[\mathbf{Q}\cdot\mathbf{u}_j(0)\right]\right\rangle \exp(-i\omega t)\mathrm{d}t\Big\} \quad (9.17)$$

We can now insert our general equation for the instantaneous displacements $\mathbf{u}_i(t)$, together with the amplitudes. The result, which is a quantum rather than classical result, is derived in Chapter 11:

$$S_1(\mathbf{Q},\omega) = \sum_\nu \Big\{ \frac{\hbar}{2\omega(\mathbf{k},\nu)} \left|F_\nu(\mathbf{Q})\right|^2$$
$$\times\left(\left[1+n(\omega)\right]\delta\left(\omega+\omega(\mathbf{k},\nu)\right) + n(\omega)\delta\left(\omega-\omega(\mathbf{k},\nu)\right)\right)\Big\} \quad (9.18)$$

where $\mathbf{Q} = \mathbf{k} + \mathbf{G}$ ($\mathbf{G}$ is a reciprocal lattice vector), $\mathbf{k}$ is a wave vector in the first Brillouin zone (note the conservation laws discussed later in this chapter), $n(\omega)$ is the Bose–Einstein factor, and $F_\nu(\mathbf{Q})$ is the *structure factor* for the ν-th normal mode:

$$F_\nu(\mathbf{Q}) = \sum_j \frac{b_j}{m_j} \exp\left(-W_j\right) \exp\left(i\mathbf{Q} \cdot \mathbf{R}_j\right) \mathbf{Q} \cdot \mathbf{e}(j, \mathbf{k}, \nu) \tag{9.19}$$

This should not be confused with the intermediate scattering function of equation (9.3) or the Bragg structure factor of equation (E.2). The use of the mode subscript (ν) will avoid confusion. Popular usage has dictated the adoption of the same symbol for the different functions, but it is clear that their definitions are distinct though similar.

There are a number of points to note from equations (9.18) and (9.19). The delta functions in equation (9.18) occur because the scattering involves only one phonon, so that the scattering function will only be non-zero for values of the phonon mode frequencies, $\omega = \pm\omega(\mathbf{k}, \nu)$. Equation (9.18) shows that there is always a greater probability of scattering for $\omega > 0$, corresponding to the creation of a phonon, rather than for $\omega < 0$, corresponding to absorption of a phonon. This is to be expected, in that at very low temperatures there is only a very small number of thermally excited phonons with $\omega > 0$, and therefore very little chance of a neutron absorbing a phonon, whereas it is always possible for the neutrons to lose energy by creation of phonons. At high temperatures the difference between the intensities for the two processes becomes smaller. In the high-temperature limit, as defined by equation (4.16), the scattering factor for any mode ν tends towards the limiting form:

$$S_1(\mathbf{Q}, \omega) \propto \frac{k_B T}{\omega^2} \delta\left(\omega - \omega(\mathbf{k}, \nu)\right) \tag{9.20}$$

We note that the probability for inelastic scattering increases with temperature. The dependence on ω^{-2} means that it is much harder to measure high-frequency modes than low-frequency modes. Accordingly high-frequency modes fall more in the domain of spectroscopy, as described in Chapter 10.

Equation (9.19) determines the intensity of single-phonon scattering, and can be exploited to provide information about the atomic motions associated with any normal mode. The first important factor is the term $\mathbf{Q}\cdot\mathbf{e}$, where we recall that $\mathbf{e}$ is a vector that lies along the direction of motion of the atom. This factor means that the intensity of one-phonon scattering will be greatest if $\mathbf{Q}$ is nearly parallel to the direction of atomic motion, and will be weakest if $\mathbf{Q}$ is nearly perpendicular to $\mathbf{e}$. Thus for measurements of a longitudinal mode, where **e** is parallel to **k**, the instrument should be set with **Q** parallel to **k**. On the other hand, **Q** has to be set nearly perpendicular to **k** for measurements of a transverse mode. The second important factor in equation (9.19) is the phase factor exp ($i\mathbf{Q}\cdot\mathbf{R}$). The symmetry of the crystal will therefore cause some modes to have zero intensity at certain values of **Q**, similar to the systematic absences of crystallography. These conditions on the observations of modes are called *selection rules*. Both factors prove to be extremely useful in performing neutron scattering measurements of phonon dispersion curves, as appropriate choices of **Q** can allow many branches to be assigned unambiguously.

Higher-order terms: multiphonon scattering

We will not discuss the higher-order terms ($m > 1$) in equation (9.13) in any detail. Each term corresponds to scattering processes involving m phonons, and the scattering function has only a weak structure in the frequency domain so that these higher-order terms will usually contribute only to the background scattering. There are no selection rules. The conservation laws as outlined below allow all modes to contribute to the multiphonon scattering for any **Q**.

Conservation laws for one-phonon neutron scattering

Consider again the scattering processes shown in Figure 9.1. The energies of the incident and scattered beams are equal to:

$$E_i = \frac{\hbar^2 k_i^2}{2m} \; ; \; E_f = \frac{\hbar^2 k_f^2}{2m} \tag{9.21}$$

respectively. The conservation law for one-phonon scattering is therefore:

$$\hbar\omega(\mathbf{Q}) = E_i - E_f = \frac{\hbar^2}{2m}\left(k_i^2 - k_f^2\right) \tag{9.22}$$

If we keep E_f constant (as is often the case; see below), and we recall that:

$$\mathbf{Q} = \mathbf{k}_i - \mathbf{k}_f \tag{9.23}$$

we can substitute for $\mathbf{k}_f$:

$$\hbar\omega(\mathbf{Q}) = \frac{\hbar^2}{2m}\left(Q^2 - 2\mathbf{Q}\cdot\mathbf{k}_i\right) \tag{9.24}$$

For each value of (ω, **Q**) the experimentalist has two variables, namely the magnitude and direction of the incident wave vector $\mathbf{k}_i$. Note that ω can be pos-

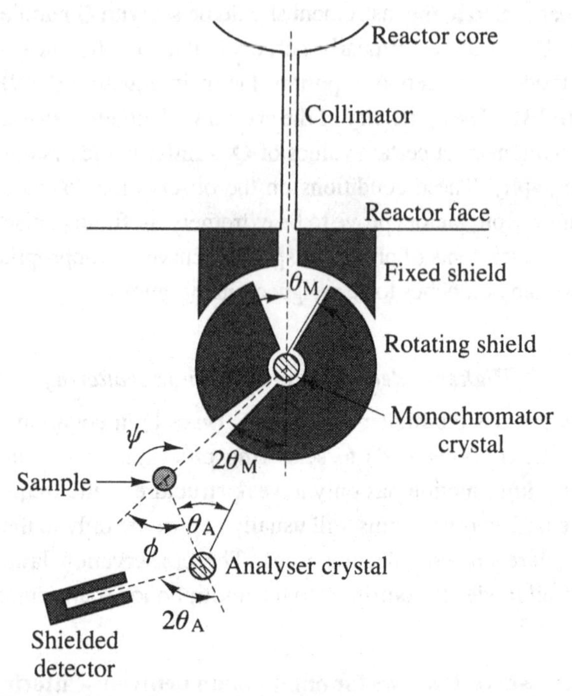

Figure 9.2: Plan of a triple-axis neutron spectrometer.

itive or negative in the application of equation (9.24). Similar considerations apply if we keep E_i constant instead of E_f. In either case the experimentalist will scan in the space defined by (ω, **Q**), and will observe peaks in intensity corresponding to phonons that are matched at any point.

Experimental inelastic neutron scattering

Since the 1960s the most successful inelastic neutron scattering technique for the measurement of phonons (and magnetic excitations) has been the use of the *triple-axis spectrometer* (TAS), which works on steady-state reactor sources (Brockhouse 1961). Other techniques were developed, but the versatility and adaptability of the fundamental design of the TAS has meant that it has become the almost universal instrument of choice for lattice dynamics measurements. As we remarked in the introduction to this chapter, most present neutron sources are nuclear reactors. However, it is quite likely that a significant number of the new sources to be developed will instead be spallation sources. The TAS cannot immediately be transferred to a spallation source, although it is hoped that a TAS-lookalike might eventually be developed for pulsed sources. At the present time the most promising instruments are indirect-geometry time-of-flight spectrometers. Both techniques are now described.

Triple-axis spectrometer

An example of the design of a TAS is shown in Figure 9.2. As described by equations (9.21) to (9.24), there are a number of variables in a scattering process, which the TAS is able to control. The first of the three axes on the TAS gives the orientation of the *monochromator*, which selects the incident wave vector ($\mathbf{k}_i$) by normal Bragg scattering. The monochromator is often made of a single crystal of a material such as silicon or germanium, or polycrystalline graphite with preferred orientation. The monochromator crystal is usually so close to the reactor that it is generally shielded with a huge drum of protective material. The second axis gives the orientation of the crystal sample itself and the angle subtended by the incoming neutron beam and the scattered neutrons that are to be detected. The third axis gives the angle of a second monochromator crystal, known as the *analyser*. This scatters neutrons of only one particular final wave vector ($\mathbf{k}_f$) into the detector. All these axes are perpendicular to the scattering plane. All angles are variable, so that the second axis moves in an arc about the first axis, and the third axis moves in an arc about the second axis.

The most common method of operation is to preselect the scattering vector $\mathbf{Q}$ relative to the crystal, so the energy is measured for a phonon of a predetermined wave vector.[3] The intensity of the scattered neutrons is measured as a function of the energy transfer; we expect a sharp peak in the resultant spectrum at an energy transfer corresponding to the phonon energy when we fulfil the conservation requirements of equations (9.22) and (9.23).

In addition to working at constant-$\mathbf{Q}$, one practice is to fix the analyser angles θ_A and $2\theta_A$ at preset values, so that E_f will be constant. This means that only neutrons of one wavelength (therefore one energy) will be Bragg scattered by the analyser crystal into the detector. Measurements performed at different values of E_f will allow some control of the resolution of the instrument; the same is true if instead E_i is kept fixed for a single measurement. Having constrained the values of $\mathbf{Q}$ and E_f (hence k_f), the only free variables that remain are the orientation of the crystal, and the incident neutron wavelength, which is changed by rotating the monochromator. This is in fact a somewhat tricky procedure as it involves rotating the shielding, which is in the

[3] There may be cases when it is preferable to preselect the energy transfer and scan values of $\mathbf{Q}$; an example is when the frequency of the phonon rises steeply with wave vector so that resolution considerations mean that the phonon frequency will not be measured very accurately with a constant-$\mathbf{Q}$ scan.

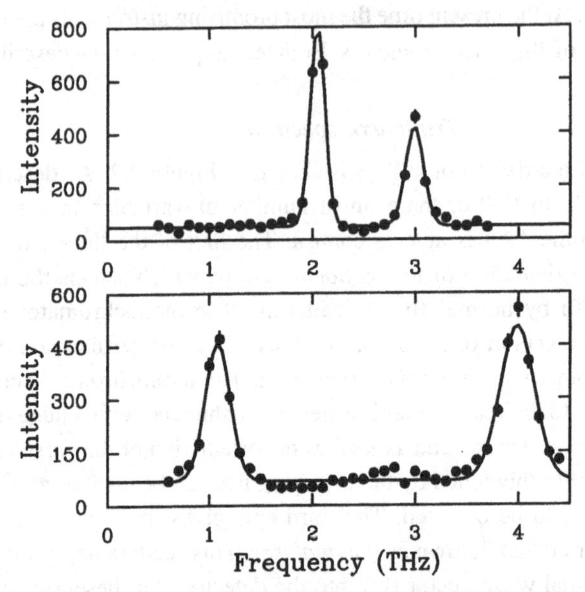

Figure 9.3: Measurements of phonon peaks in calcite (Dove et al. 1992b), obtained, using a triple-axis spectrometer operating in constant-$\mathbf{Q}$ mode. The curves are Gaussians that have been fitted to the data. The top graph is for $\mathbf{Q} = (2.225, 0, 3.1)$ and the bottom graph is for $\mathbf{Q} = (0.5, 0, 10)$.

form of a huge drum that is several feet in diameter and may weigh several tonnes! The crystal and the analyser assembly have to be rotated together in order to keep $\mathbf{Q}$ constant. For fixed $\mathbf{k}_f$ and $\mathbf{Q}$, all we have to vary are ϕ (the angle through which the neutron beam is scattered) and ψ (the orientation of the crystal). As might be expected, the apparatus is completely controlled by computer.

Some examples of results from experiments to measure phonon dispersion curves in calcite are shown in Figure 9.3, and the corresponding measured dispersion curves are given in Figure 9.4. In practice the phonon peaks are not infinitely sharp as there is a finite resolution of the spectrometer (which is chosen as a compromise by the experimentalist; too high a resolution will lead to a loss in intensity). Also the peaks can be broadened as a result of anharmonic interactions that lead to a finite lifetime for the phonon, as discussed in Chapter 8. The examples given in Figure 9.3 show quite clear phonon peaks. In many cases the peaks may overlap.

For a general scattering vector $\mathbf{Q}$ there will in principle be the same number of peaks as modes, which for complex crystals will give rise to very complicated spectra. It is possible to interpret these spectra using the structure factor of equation (9.19). Before the experiment is performed, a postulated model for

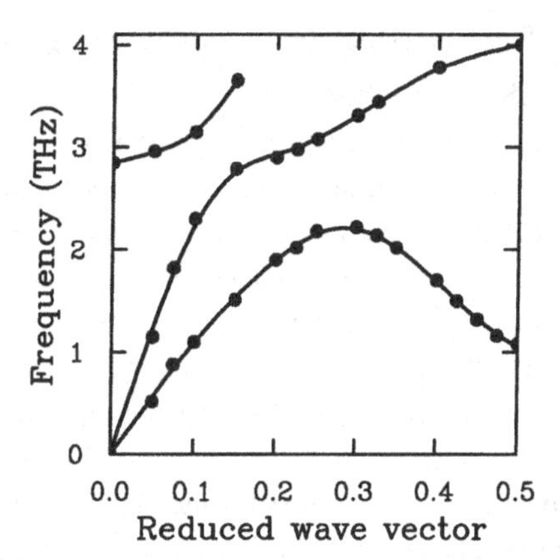

Figure 9.4: Dispersion curves of calcite for wave vectors along $[1, 0, \bar{4}]$ as determined by neutron scattering measurements on a triple-axis spectrometer (Dove et al. 1992b). The data shown in Figure 9.3 are for the reduced wave vectors of 0.225 and 0.5.

the force constants of the crystal is used in a calculation of the dispersion curves and mode eigenvectors. These pre-calculated mode eigenvectors for a wide range of values of **Q** are used to calculate the neutron scattering intensity for each mode. Some intensities will be calculated to be strong, and others will be calculated to be weak. The experimentalist then chooses values of **Q** such that only a few modes are expected to be strong. This makes it possible to identify the observed modes with the calculated modes, provided that the original model is reasonable. An experiment is never as straightforward as this sounds, and sometimes the model only works well for high-symmetry wave vectors (such as the centre and boundaries of the Brillouin zone), but there is usually enough information to enable most of the observed modes to be interpreted, albeit with a bit of thought.

With high-quality data it may be possible to extract the mode eigenvectors for complex crystals, namely those for which the eigenvector is not determined only by symmetry, by fitting data from a range of values of **Q** (but of course with the same phonon wave vector **k**) against equation (9.19). This is usually carried out only for special wave vectors, such as particular zone boundary points. Two examples are the determination of the set of mode eigenvectors at a zone boundary wave vector in the low-symmetry crystal of naphthalene, where the structure factor was fitted to constant-**Q** measurements at 24 values of **Q** (Pawley et al. 1980), and a zone boundary mode in quartz which was observed to have a strong temperature dependence (Boysen et al. 1980). Chaplot et al. (1981) have also determined the modes of motion associated with the three lowest-frequency internal modes of the molecular crystal anthracene from inelastic neutron scattering intensities.

Time-of-flight spectrometers using indirect geometry

Experiments on the TAS are performed using measurements of scattering angles to determine the energy and wave vector transfers, collecting data for one ($\mathbf{Q}$, ω) value at a time. This type of measurement, and the use of Bragg reflection to select the energies, exploits the high flux of a steady-state reactor as integrated over time. The integrated neutron flux from a spallation source is much lower so that the use of a TAS would be impractical. However, a high "effective flux" can be obtained by exploiting the time structure of the neutron pulses produced with a spallation source. For example, at the ISIS spallation source (UK) pulses of neutrons with time width of 20 μs are produced every 20 ms. The narrow pulse width and long time between pulses allows the incident energy of a neutron to be obtained from the measurement of its flight time. The flight time τ for which a neutron of energy E and wave vector k travels the distance L is given by

$$\tau = \frac{mL}{\hbar k} = L\left(\frac{m}{2E}\right)^{1/2} \tag{9.25}$$

If $L = 10$ m, τ will be in the range 1–10 ms for neutrons with energy in the range 5–500 meV. Since the initial pulse width is narrow compared with this time, it is possible to separate the different incident neutron energies by their flight times and to use them all simultaneously in a measurement. This approach has so far been used only in time-of-flight spectrometer designs based on the principle of *indirect geometry*.

Consider the diagram of a time-of-flight spectrometer shown in Figure 9.5 (Steigenberger et al. 1991). The crystal orientation and analyser angle are kept fixed throughout the experiment, which means that $\mathbf{k}_f$ is constant. Thus the time taken to travel from the crystal to the detector (τ_f) is known exactly. Knowing by measurement the total time taken for the neutron to travel from the target to the detector, subtraction of τ_f gives the time taken for the neutrons to travel from the target to the crystal. This therefore gives the incoming neutron wave vector $\mathbf{k}_i$. Across the whole pulse we have a wide range of values of $\mathbf{k}_i$ and hence of $\mathbf{Q}$. But given that the directions of both $\mathbf{k}_i$ and $\mathbf{k}_f$ are fixed, the locus of the values of $\mathbf{Q}$ in reciprocal space is a straight line; this is illustrated in Figure 9.6. From equations (9.22) and (9.23) we see that by measuring the intensity of the scattered neutrons as a function of flight time we are actually measuring the intensity as a function of a parabolic locus in ($\mathbf{Q}$, ω) space, rather than as a function of only $\mathbf{Q}$ or ω as is common with a TAS. In order to improve the scope of measurements, several analyser–detector systems are used which collect data simultaneously. For example, the PRISMA spectrometer at ISIS has 16 detectors. It has been shown (Steigenberger et al. 1991) that if each analyser and detector is set with a constant

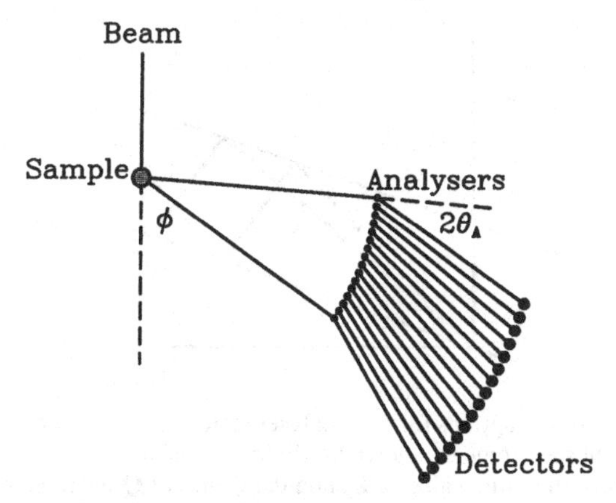

Figure 9.5: Diagram of a time-of-flight indirect-geometry spectrometer with 16 detectors (Steigenberger et al. 1991).

value of the ratio sin ϕ / sin θ_A, with the angles defined in Figure 9.5, each analyser–detector system gives measurements of different (**Q**, ω) trajectories for the same locus of values of **Q**. The advantage is that with the use of several detectors the spectrometer measures phonons over an area of (**Q**, ω) space during a single run, whereas a large number of measurements using different spectrometer configurations are required with a TAS.

An example of a single detector spectrum from an experiment on calcite (Dove et al. 1992b) is shown in Figure 9.7, and the corresponding set of dispersion curves measured in several different Brillouin zones using just three settings of the instrument is shown in Figure 9.8. The advantage of such a wide coverage of (**Q**, ω) space within a single measurement, which clearly enables phonon dispersion curves to be mapped out in a relatively short time, is offset by limitations on the flexibility to perform a range of measurements and to vary the resolution. However, such instruments are sufficiently new that the means to overcome these limitations may not be long in coming, enabling the full potential of these spectrometers to be realised.

Advantages of neutron scattering and some problems with the technique

1. In principle, *all* phonon branches can be measured by neutron scattering, provided that the Brillouin zone for the measurement is chosen with large enough structure factor.
2. With neutron scattering we can measure phonon frequencies for all phonon wave vectors. Neutron scattering is the only available technique for measuring complete dispersion curves, and is the only technique for measuring phonons with wave vectors other than $\mathbf{k} \sim 0$.

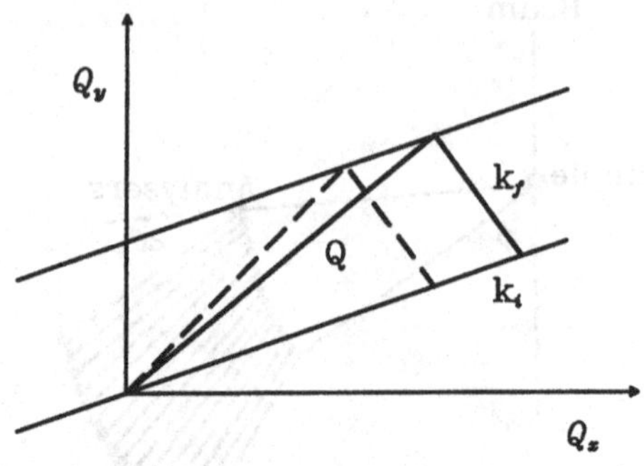

Figure 9.6: Scattering diagrams for two analyser–detector systems on an indirect-geometry time-of-flight spectrometer, shown as a bold line and as a dashed line. Both scattering diagrams have the same value of $\mathbf{k}_f$, and the values of $\mathbf{Q}$ lie along one given direction in reciprocal space (Steigenberger et al. 1991).

3 It is fairly easy to control the sample environment. Neutron scattering experiments are usually fairly large in scale, which greatly assists in measuring temperature or applying external fields. Neutrons are only weakly absorbed by the walls of sample chambers, which makes the apparatus for control of the sample environment easier to construct.

4 Because neutron beams are relatively weak compared with X-ray or laser sources, large single crystals are usually needed for inelastic neutron scattering measurements. Typical sample sizes are of the order of 1 cm^3 or larger, but with high-intensity neutron beams and instruments designed to optimise the scattered intensity and reduce the background scattering it is possible to use crystals as small as a few mm^3. Even crystals of this size may be difficult or even impossible to obtain, either because the crystals cannot be grown or because of twinning associated with phase transitions.

5 Neutron scattering can be performed only at central facilities, rather than in the local laboratory. The advantage of a laboratory-based facility is that it can be accessed at all times, and experiments can be performed at the experimentalists' convenience (or in immediate response to their impatience!). Most neutron scatterers need to book beam time on a central facility some time in advance of the experiment. However, for many the travelling that is required, which may mean travelling abroad, is not seen as a disadvantage!

6 Typical neutron beam intensities are much lower than the intensities from X-ray or light sources, which means that the time for a typical neutron scattering measurement is much longer than for other techniques. The consequence of this is that other experimental methods are much better if measurements are required over a wide range of external conditions. Generally,

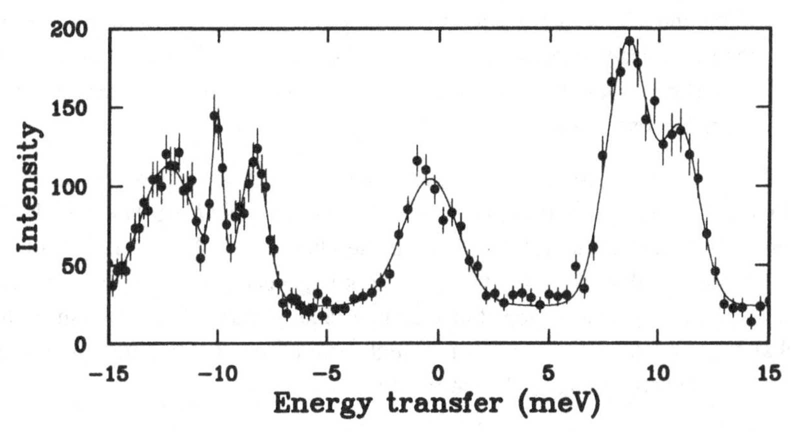

Figure 9.7: The spectrum from one detector obtained on the indirect-geometry time-of-flight spectrometer PRISMA. The sample was calcite aligned so that the wave vectors of the phonons are along $[1, 0, \overline{4}]$ (Dove et al. 1992b).

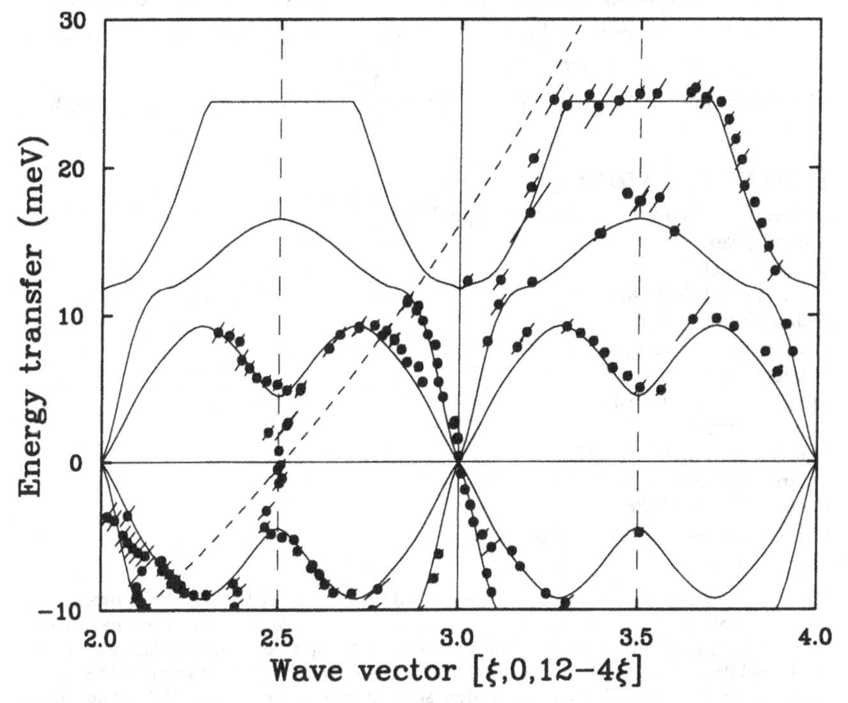

Figure 9.8: A set of dispersion curves measured in calcite using just three settings of the time-of-flight spectrometer (Dove et al. 1992b). The data are shown across several Brillouin zones. This figure can be compared with the corresponding measurements obtained using a triple-axis spectrometer shown in Figure 9.3.

if the interest is in the effects of changing conditions (e.g. temperature), then the disadvantages of other methods such as light scattering are less important, unless the interesting phenomena occur at wave vectors away from the zone centre.

It has recently been demonstrated that phonons can also be measured directly by inelastic X-ray scattering[4] (Dorner et al. 1987; Burkel et al. 1989, 1991; Hofmann et al. 1992). Because the changes in photon energy are so much smaller than the energy of the incoming beam, these experiments are technically very demanding. For example, energy changes are measured by changing the lattice spacing of a high-quality monochromator, which is accomplished by changing the temperature.

Summary

1 The properties of neutrons have been shown to make neutron scattering an ideal tool for the measurement of phonon dispersion curves.
2 The general formalism for neutron scattering has been developed to provide an expression for the scattering of neutrons involving either creation of a single phonon or absorption of a single phonon.
3 Two experimental techniques of neutron scattering have been described.

FURTHER READING

Ashcroft and Mermin (1976) ch. 24; app. N, O
Bacon (1986)
Bée (1988) ch. 2
Blinc and Zeks (1974)
Brüesch (1986) ch. 6
Cochran (1973) ch. 5, 8, 10
Dorner (1982)
Iqbal and Owens (1984) ch. 5
Lovesey (1984)
Marshall and Lovesey (1971)
Squires (1978)
Skold and Price (1986)
Willis and Pryor (1975) ch. 7–8

[4] Before neutron scattering methods were developed (Brockhouse and Stewart 1955; Brockhouse 1961) a number of efforts were made to deduce dispersion curves indirectly from X-ray thermal diffuse scattering. The intensity of the diffuse scattering can be expressed in terms of lattice dynamics (Willis and Pryor 1975, p 215; Ashcroft and Mermin 1976, pp 794–795), assuming that the dominant contribution to the diffuse intensity is from single-phonon scattering. One of the more important studies of this type was the determination of the dispersion curves for aluminium by Walker (1956). As discussed by Willis and Pryor (1975, pp 215–218), there are a number of experimental problems with this approach, including the existence of multi-phonon scattering, which lead to systematic errors in the determination of the dispersion curves that cannot be reliably estimated.

10

Infrared and Raman spectroscopy

Infrared spectroscopy and Raman scattering provide methods of measuring phonon frequencies that are complementary to neutron scattering. We show that the use of electromagnetic radiation leads to constraints that do not exist for neutron scattering, but which can be exploited. In this chapter we give an introductory description of the background theory and the experimental methods. We then outline some of the areas of application.

Introduction

Neutrons are complemented as probes of the dynamic behaviour of crystals by the quanta of electromagnetic radiation, photons. Having zero rest mass, photons do not scatter from phonons in the same way that neutrons do; the differences are highlighted by considering photon scattering from the viewpoint of neutron scattering.[1] We note from the start the numerical values of the important quantities. The photon angular frequency ω and wave vector k are related by $\omega = ck$, where c is the velocity of light (3×10^8 m s^{-1}). This linear relation contrasts with the quadratic relation for neutrons, equation (9.1). A change in energy of the photon due to the absorption or creation of a phonon will therefore cause a linear change in its wave vector. If the change in frequency is 5 THz, say, for a typical optic mode, the corresponding change in wave vector $\Delta k/2\pi$ will be $\frac{1}{6} \times 10^{-5}$ Å^{-1}. This corresponds to a phonon wave vector that is very close to the Brillouin zone centre in a crystal of typical unit cell dimensions. These numbers demonstrate that photons will be scattered or absorbed only by phonons with very long wavelength.

We will consider first the case where the photon is absorbed to create a phonon of the same frequency and wave vector. Such an experiment will

[1] Not a viewpoint necessarily favoured by spectroscopists!

involve shining a polychromatic beam of radiation on the sample, and measuring the frequencies at which absorption occurs. These will typically be in the infrared region of the electromagnetic spectrum, hence this technique is known as *infrared spectroscopy.*[2] Even when the conditions of energy and wave vector conservation are met, there may be symmetry factors that prevent absorption; in the language of scattering theory we would say that the *cross section* or *scattering* (*structure*) *factor* is zero, and in the language of spectroscopy we would say that the *selection rules* do not allow the absorption process.

We next consider a true scattering process, in which the photon is scattered with a change in frequency. In this case it is common to perform experiments using monochromatic radiation, usually lasers in the visible region of the electromagnetic spectrum. If we take an optic mode with a frequency ω_0 that has a negligible dependence on wave vector, and consider the usual case where the light beam is scattered through an angle of 90°, the scattering equations are:

$$\omega_f = \omega_i \pm \omega_0 \tag{10.1}$$

$$|\mathbf{Q}|^2 = \left|\mathbf{k}_i \pm \mathbf{k}_f\right|^2 = k_i^2 + k_f^2 = c^2\left(\omega_i^2 + \omega_f^2\right) \tag{10.2}$$

As with neutron scattering, photons can be scattered either by absorbing or creating a single phonon. This type of scattering is called *Raman scattering,*[3] named after one of its experimental discoverers.[4] Similar scattering processes can also occur that involve acoustic phonons; in this case the changes in the frequency of the light beam are so small that different experimental techniques are required. The scattering of photons by acoustic modes is called *Brillouin scattering*, in this case named not in honour of the experimental discovery of the effect but after the theoretical prediction[5] (Brillouin 1914, 1922).

We noted in the previous chapter that neutrons can be scattered by more than one phonon, but the conservation laws for multiphonon scattering are sufficiently slack that the scattering is not constrained to give peaks. Multiphonon scattering processes are also allowed in spectroscopy. However,

[2] The early measurements include those of Rubens and Hollnagel (1910) and Barnes (1932).
[3] Technically, when a single phonon is involved it is known as *first-order Raman scattering*.
[4] Raman and his co-workers (Raman 1928; Raman and Krishnan 1928a,b) were initially mainly interested in Raman scattering from molecular vibrations. Raman scattering was observed in quartz by Landsberg and Mandelstam (1928, 1929) at around the same time.
[5] Brillouin scattering was first observed by Gross (1932a–d), who recognised that this was different from normal Raman scattering and took his observations as confirmation of the Debye model for the elastic waves of a crystal.

because the only phonons involved are those with small wave vector, the multiphonon processes still lead to sharp peaks in the measured spectra. These are often called *combination* or *overtone* bands.

The interaction between photons and phonons, whether by scattering or absorption, can be treated in a number of ways. The macroscopic approach begins with Maxwell's equations of electromagnetism as applied to a dielectric medium, and connects with the normal modes at the stage when the response of the crystal to the electric field of the photon is required. The microscopic approach is to consider scattering and absorption processes as involving transitions between different quantum states, and to calculate the appropriate matrix elements in a perturbation treatment. Both approaches require a considerable amount of background theory if one is to avoid merely quoting results; instead we adopt a more phenomenological approach to the understanding of the different processes.

Vibrational spectroscopy by infrared absorption

Fundamental principles

One can view the absorption of infrared radiation as an analogy to forced harmonic motion in a macroscopic mechanical system. The long wavelength infrared photon will cause a local dielectric polarisation of the crystal which will oscillate in time. If the crystal cannot easily respond to this forced vibration, the radiation will not be absorbed. However, absorption will be strong when the frequency of the radiation is the same as that of a phonon frequency, provided that the eigenvector of the phonon gives rise to the formation or change of the local dipole moment. In this case, the phonon frequency corresponds to a resonance frequency, and absorption of the infrared radiation is significantly increased. In general the absorption $A(\omega)$ will be proportional to the power spectrum of the fluctuations of the dipole moment M generated by the relevant normal mode:

$$A(\omega) = \int \langle M(0)M(t) \rangle \exp(-i\omega t)\mathrm{d}t \qquad (10.3)$$

The instantaneous dipole moment can be expressed as a linear combination of all the phonon modes that give rise to a polarisation:

$$M(t) = \sum_j m^{-1/2}(j) q_j \mathbf{e}(j, \mathbf{k}, \nu) Q(\mathbf{k}, \nu) \qquad (10.4)$$

where q_j is the effective charge of an ion.[6] Thus the absorption is formally given as

$$A(\omega) \propto \left| \sum_j m^{-1/2}(j) q_j \mathbf{e}(j, \mathbf{k}, \nu) \right|^2 \times \int \langle Q(\mathbf{k}, \nu, t) Q(-\mathbf{k}, \nu, 0) \rangle \exp(-i\omega t) \mathrm{d}t$$

$$= \left| \sum_j m^{-1/2}(j) q_j \mathbf{e}(j, \mathbf{k}, \nu) \right|^2 \times \frac{\hbar}{2\omega(\mathbf{k}, \nu)} \left[n(\omega) + 1 \right] \delta(\omega - \omega(\mathbf{k}, \nu)) \tag{10.5}$$

This expression is similar in form to the expression for the inelastic scattering of neutrons, equation (9.17).[7] The effective charge of each atom has a role similar to that of the neutron scattering length. Unlike neutron scattering, however, infrared absorption occurs *only* for the set of normal modes that give rise to a local dipole moment.[8] These modes are said to be *infrared active*, and the set of infrared active modes for any crystal can be determined from group theory. The derivation of equation (10.4) assumed absorption by harmonic phonons. In practice anharmonic interactions will cause the phonon absorption peaks to be broadened, and this broadening can be easily measured on high-resolution spectrometers. The absorption spectrum $A(\omega)$ given by equation (10.5) should therefore be modified by convoluting with a damped oscillator spectrum. In more exact treatments of infrared absorption, the phonon lifetime is incorporated into the theory at an earlier stage.

Transmission experiments with polycrystalline samples

Infrared absorption experiments with polycrystalline samples performed by measuring the transmission through the sample are the easiest infrared absorption experiments, from the two viewpoints of the experimental methods and the data analysis. The central idea is to use a filament source for the incident infrared radiation, extending from a few cm^{-1} to a few thousand cm^{-1}, and to

[6] For a rigid (unpolarisable) ion the effective charge will be identical to the actual ionic charge. However, when polarisability is taken into account, as in a shell model, it turns out that the effective charge is not the same as the ionic charge. A more complete treatment would handle the cores and shells separately by noting that part of the induced dipole moment associated with a given normal mode arises from the ionic polarisation; equation (10.4) simply associates the induced moment with the atomic displacements.

[7] The definition of equation (10.5) differs slightly from equation (9.18) in that the sign of the frequency of the created phonon is different in the two equations. We retain this difference since only one process is involved in equation (10.5).

[8] These modes are called *polar modes*.

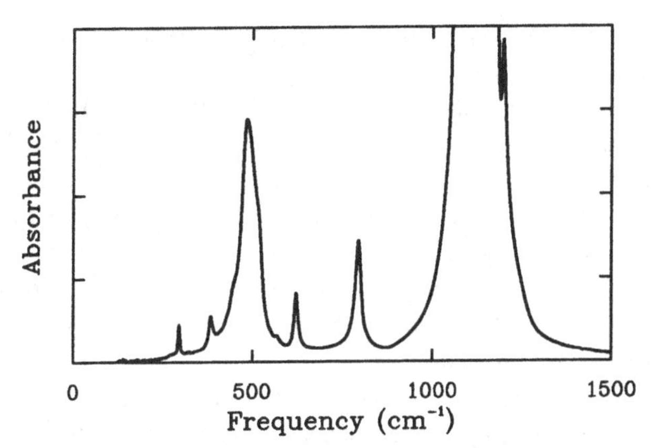

Figure 10.1: An infrared absorption spectrum for cristobalite, SiO_2, at room temperature. The peaks in the spectrum correspond to phonon absorption of the incident radiation.

measure the amount of radiation transmitted through the sample by comparing the measured spectrum with the spectrum of the source. The transmittance can then be converted to a relative absorption. The absorption spectrum will contain peaks corresponding to the phonon frequencies; a typical example is shown in Figure 10.1.

The measurements on a modern spectrometer are performed using a Fourier transform interferometer. This is a variant of a Michelson interferometer, in which one of the mirrors moves backwards and forwards, yielding a spectrum as a function of time for each traverse of the mirror. This spectrum is then Fourier transformed on a computer to produce an absorption spectrum. A large number of spectra are collected from many separate scans of the mirror, and these are all averaged to produce a final spectrum with an excellent signal-to-noise ratio. Older spectrometers use grating methods to monochromate the infrared beam, leading to a direct measurement of the intensity of the transmitted spectrum as a function of frequency. Fourier transform methods are much faster than grating methods, they produce spectra of superior quality in terms of the statistical noise and resolution, and grating methods are more susceptible to stray light sources.

As the absorption can be quite strong, very thin samples are required. It is common to produce pellets that contain a small quantity of material held in an appropriate medium. For many applications alkali halides are good materials for the sample medium, as their infrared spectra are simple and well characterised, and for many applications their peaks do not overlap the spectral regions of interest. For studies of low-frequency modes, polymers may be preferred as medium materials, as they do not have spectral peaks in the low-frequency region. Both alkali halides and polymers compact well in the preparation of the pellets.

Transmission experiments are sensitive only to the transverse optic modes. However, with the use of powders it is possible to observe some absorption from the longitudinal modes due to reflection from the surfaces of the crystallites in the sample material.[9] Depending on the difference in frequencies between the transverse and longitudinal modes, the effect of the longitudinal modes may be to produce extra peaks, or to give an asymmetric broadening of the transverse mode peaks. Generally these effects are weak, but in some cases they can cause severe problems.

Single crystal measurements can be performed either by measuring the absorption in transmission experiments, as in the measurements on polycrystalline samples described above, or by measuring the reflectance from a surface. In the latter case the infrared radiation is reflected following the excitation of the vibrations in the crystal by the incident radiation. The resultant spectrum is not easily analysed; unlike absorption methods the reflectance spectrum does not simply consist of peaks corresponding to phonon frequencies. The deconvolution of the reflectance spectrum takes us beyond the scope of this book. It should be noted that the reflectance spectrum will include contributions from both transverse *and* longitudinal modes.

Raman spectroscopy

Fundamental principles

A small fraction of a light beam that passes through any material will be scattered. Most of this will be scattered elastically, that is, with no change in frequency. This is known as *Rayleigh scattering*, and the amplitude is proportional to the polarisability α of the material, appropriate to the relevant scattering geometry. The dependence on the polarisability follows from the fact that the scattered radiation originates from the dipole oscillations (polarisation fluctuations) induced by the light wave. The amplitude of the scattered wave will be proportional to the induced moment **M**:

$$\mathbf{M} = \alpha \mathbf{E} \tag{10.6}$$

where **E** is the vector amplitude of the electric field of the incident radiation.

The polarisability at any region of the crystal will not be static, but will fluctuate as the lattice vibrations change bond lengths and contact distances

[9] Recall that for polar modes the longitudinal and transverse optic modes at $\mathbf{k} = 0$ have different frequencies, as given by the Lyddane–Sachs–Teller relation, equation (3.29).

and hence the local electron density. We can therefore write the time-dependent polarisability $\alpha(t)$ when it has been modified by a lattice vibration of frequency ω' as

$$\alpha(t) = \alpha_0 + \alpha'\left(\exp(i\omega' t) + \exp(-i\omega' t)\right)Q \qquad (10.7)$$

where α_0 is the static polarisability, α' is a constant, and Q is the amplitude of the lattice vibration. The amplitude of the scattered wave is therefore given as

$$\begin{aligned} M(t) &= E\exp(i\omega t)\alpha(t) \\ &= E\alpha_0\exp(i\omega t) + E\alpha'\left(\exp\left(i(\omega+\omega')t\right) + \exp\left(i(\omega-\omega')t\right)\right) \end{aligned} \qquad (10.8)$$

The first component of the scattered wave is the elastic Rayleigh scattering. The other two components are scattered with a change in frequency; these are the two Raman scattering processes corresponding to creation and absorption of a phonon of frequency ω'.

When these ideas are applied to the complete set of normal modes in a crystal we need to note that the constant α' must be expressed as a tensor that takes account of the directions of the incoming and scattered beams, and the polarisations of the two beams. There will be a different form of this tensor depending on the actual normal mode involved. We can therefore rewrite equation (10.7) as

$$\alpha(t) = \alpha_0 + \sum_{\mathbf{k},\nu}\alpha'(\mathbf{k},\nu)\left(\exp(i\omega(\mathbf{k},\nu)t) + \exp(-i\omega(\mathbf{k},\nu)t)\right)Q(\mathbf{k},\nu) \qquad (10.9)$$

The intensity $I(\omega)$ of the scattering will be determined by the power spectrum of $\alpha(t)$:

$$\begin{aligned} I(\omega) \propto \int\langle\alpha(0)\alpha(t)\rangle\exp(-i\omega t)\mathrm{d}t = \alpha_0^2\delta(\omega) \\ +\sum_{\mathbf{k},\nu}\gamma(\mathbf{k},\nu)\left\{\left(n(\omega,T)+1\right)\delta\left(\omega-\omega(\mathbf{k},\nu)\right) + n(\omega,T)\delta\left(\omega+\omega(\mathbf{k},\nu)\right)\right\} \end{aligned} \qquad (10.10)$$

where the first term on the right hand side of the equation is the elastic Rayleigh scattering. The coefficient $\gamma(\mathbf{k},\nu)$ subsumes the parameter α', and therefore takes account of the polarisability for the specific scattering geometry.

It is apparent that the equation for the intensity of Raman scattering is, apart from the coupling constants, very similar in form to that for inelastic neutron scattering, as one might have expected. The polarisability tensor has a similar role to the neutron scattering length, providing the coupling between the probe and the material. Unlike the case of neutron scattering, however, the relevant components of the polarisability tensor can be zero for particular phonon modes irrespective of the relative orientations of the scattering geometry and crystal. This is a symmetry property, and a group-theoretical analysis of the crystal symmetry will indicate which modes are allowed to take part in Raman scattering – these modes are said to be *Raman active*. Symmetry arguments cannot, however, provide information on the strength of the Raman scattering, that is, on the size of the components of the polarisability tensor. This is determined by the electronic structure of the crystal.

Raman scattering involving creation of a phonon is called the *Stokes process*, and scattering involving absorption of a phonon is called the *anti-Stokes process*. As for neutron scattering, the processes of phonon creation and absorption are weighted by different thermal factors, $n(\omega, T) + 1$ and $n(\omega, T)$ respectively.

Experimental techniques

The principles of the experimental methods of Raman scattering are straightforward. The source of the monochromatic light beam is a laser. The light is usually scattered through an angle of 90° and measured using a photomultiplier. The spectrum is obtained by using gratings to select and change the measured frequency. The resolution can be controlled by apertures placed in the incoming and scattered beams. Although grating methods are the most common methods in use, Fourier transform methods similar to those used for infrared spectroscopy have recently been developed for Raman spectroscopy.

Raman scattering experiments can be performed using both polycrystalline and single crystal samples. If powders are used, the samples are usually encapsulated in thin glass tubes. Single crystals are used when the symmetries of the phonons are to be determined, because the selection rules operate differently for different crystal orientations and different polarisations of the incoming and scattered beams.

A typical Raman scattering spectrum from a polycrystalline sample is shown in Figure 10.2.

Brillouin spectroscopy

Brillouin scattering is, in principle, similar to Raman spectroscopy, except that it is used for measuring the acoustic modes rather than the optic modes. Using the methods of Chapter 7 it is possible to express the normal mode coordinates

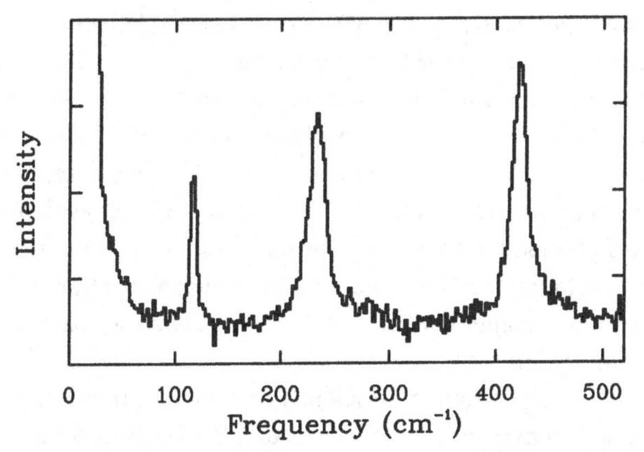

Figure 10.2: Raman scattering spectrum from cristobalite at room temperature. The peaks correspond to the frequencies of the phonons that are created by the incident radiation. The large peak at zero frequency is the elastic Rayleigh peak.

in terms of strain coordinates, and therefore derive the scattering cross section in terms of the dependence of the polarisability on strain.

The essential experimental difference between Raman and Brillouin spectroscopy is that the changes in frequency for the scattered beam are much smaller in Brillouin scattering. Therefore a more sensitive measurement of the frequency of the scattered beam is required, for which it is usual to use a Fabry–Perot interferometer. Moreover, it is important to know the wave vector transfer accurately, which is not the case for Raman and infrared spectroscopy. Hence it is essential to have an accurate alignment of the single crystal, and a good knowledge of the refractive indices in order to correct for retardation of the light beam within the crystal.

Advantages and disadvantages of spectroscopy

If one considers only the information content, the restriction of spectroscopic methods to measurements of only long-wavelength phonons and the stringent selection rules would appear to make neutron scattering a preferred technique. However, there are a number of experimental advantages that in many cases mean that spectroscopy is a more useful technique than neutron scattering:[10]

1 Spectroscopy gives very good frequency resolution, typically of the order of 1–2 cm^{-1}. Typical resolution for neutron scattering is several times larger. There is always a compromise between resolution and intensity of

[10] The conclusion of the comparison of the different techniques is that you choose the best technique for the information you require!

the measured spectra. Because neutron sources are relatively weak, greater importance has to be attached to intensity rather than resolution. Intensity is rarely a problem with spectroscopic techniques, so that the greater importance can be placed on the resolution. Good resolution is important for accurate measurements of phonon frequencies, and essential if small changes in peak frequencies are to be measured. Good resolution is also important if phonon lifetimes are to be measured. Generally the linewidth of a phonon is larger than the resolution of a Raman or infrared absorption spectrometer but smaller than the resolution of a neutron spectrometer in a typical configuration.

2 The intensity of a spectrum is sufficiently high that the time taken to obtain a complete spectrum is not very long; a matter of minutes rather than hours as in a neutron scattering experiment. This means that measurements of phonon peak frequencies, intensities or lifetimes can be measured as detailed functions of external conditions such as temperature or pressure. Spectroscopy is better suited than neutron scattering for this kind of study, provided that the mode of interest can be observed by spectroscopic methods.

3 The intensity of a spectrum is also so high that it is quite easy to measure high-frequency modes. These modes are effectively inaccessible to most neutron scattering instruments because the intensity will be too low. Thus high-frequency modes will generally be better measured using infrared or Raman spectroscopy.[11] The restriction on measurements to small wave vector is actually of less concern for these modes, for it is generally found that the high-frequency modes have only a weak dependence on wave vector.

4 Given that small single crystals or powders can be used in both infrared absorption and Raman spectroscopy, the techniques can be used for many materials.[12] In general neutron scattering methods are restricted to the use of large single crystals, which are frequently not available.[13]

[11] In favourable circumstances it is still possible to measure high-frequency modes using neutron scattering. The instrumental requirement is that the neutron beam should come from a hot moderator, so that the maximum in the neutron flux is at a high energy. Neutron spectroscopy is easier if the scattering is from hydrogenous samples, as hydrogen has a particularly large cross section for incoherent scattering.

[12] There will be problems if the samples are completely opaque to the incoming light beam, or if they fluoresce.

[13] It is in principle possible to do neutron spectroscopy on polycrystalline samples. However, since neutron scattering is not restricted to only phonons with small wave vectors, the resultant spectra will contain peaks associated with phonons of wave vectors throughout the Brillouin zone. The spectra will therefore contain so many overlapping peaks that they will be difficult to interpret. However, it is possible to perform careful experiments that give a correct averaging over all scattering vectors to allow measurements of the phonon density of states weighted by the different scattering lengths. Alternatively, if dispersion is small, as it often is for high-frequency modes, it may be possible to extract useful results from neutron spectroscopy from polycrystalline samples; see footnote 11.

5 The sample environment can be easily controlled in both neutron scattering and spectroscopic experiments. One advantage of the use of small samples in spectroscopy is that it is possible to use diamond anvil cells to attain much higher pressures than can be obtained on the larger hydrostatic pressure cells designed for neutron scattering experiments. The only technical difficulty with optical spectroscopy is that of finding optical windows that will suit either high or low temperatures and high pressures. For high-temperature work there is a problem with optical spectroscopy associated with high background signals due to the black-body radiation from the sample and furnace.

The major disadvantage of spectroscopy is the existence of the selection rules, which means that to obtain as complete a picture as possible both Raman and infrared methods need to be used. However, there will frequently be some modes that are invisible to both techniques, and which can therefore be measured only by neutron scattering. The positive aspect to the selection rules is that they enable the symmetries of the modes to be determined. Moreover, in some applications described below, the selection rules can be used to provide information about the symmetry of local environments associated with short-range order. The other disadvantage is that Raman and infrared spectroscopy can measure only phonons with small wave vector, although the positive aspect of this is that this enables the use of polycrystalline samples.

Qualitative applications of infrared and Raman spectroscopy

The set of vibrations for any material will be more-or-less unique to that particular material. This property gives vibrational spectroscopy a valuable role for qualitative identification methods, both for crystal and molecular identification. In this sense the vibrational spectra represent fingerprints for the phases being identified. Because of the relative simplicity of the experimental methods, vibrational spectroscopy is a relatively convenient and rapid method for phase identification, which may be required after chemical treatment or following large changes in sample environment. Examples are given in McMillan and Hofmeister (1988).

Identification methods can also be applied using specific features in the spectra rather than the total spectra. For example, the vibrational spectrum of water is significantly different from the single-line spectrum of the hydroxyl OH^- ion. The presence of either spectrum in the vibrational spectrum of a hydrated mineral will give unambiguous information concerning the chemical behaviour of the water molecules within the mineral (Rossman 1988). Similar methods can be used in the structural studies of silicate glasses. The vibrational spectra of SiO_4 and SiO_6 units are sufficiently different that by observing the internal modes of these units it is possible to deduce the presence or absence of these coordination polyhedra. Moreover, different connectivities of SiO_4 tetrahedra will give rise to different vibrational frequencies; for example, a dangling Si–O bond will have a different frequency for the bond-stretching vibration than if the oxygen atom is bonded to two silicon atoms.

Quantitative applications of infrared and Raman spectroscopy

Accurate measurements of phonon frequencies

The high resolution of spectroscopic measurements means that accurate measurements can be obtained for phonon frequencies. Although the low-frequency dispersion curves are better measured using neutron scattering, the high-frequency modes, as we have pointed out, are usually more easily measured by spectroscopy. All the modes are important for the development of an interatomic potential model. For example, in the recent development of empirical models for calcite (Dove et al. 1992c) and PbI_2 (Winkler et al. 1991b), both the low-frequency modes measured by neutron scattering and the high-frequency modes measured by Raman and infrared spectroscopy, as well as elastic constant measurements, were included in the set of experimental data against which the parameters of the model were fitted.

A key aspect of the measurement of phonon frequencies by spectroscopy is the assignment of each peak to a mode of a given symmetry. Group-theoretical analysis will give the expected numbers of modes for each symmetry, and which modes are active in both Raman and infrared spectroscopy. Furthermore, if single crystals are studied using polarised beams, group theory is also able to tell which modes should appear in a spectrum for any geometric configuration. This often enables all modes to be unambiguously assigned to their symmetry representations. If polycrystalline samples have to be used, the process of assignment is more tricky, but can be helped if a realistic set of calculated mode frequencies is already available.

Brillouin spectroscopy, on the other hand, is primarily used for the determination of elastic constants. The alternative experimental methods are ultrasound techniques and neutron scattering. Brillouin spectroscopy will usually give more accurate results than neutron scattering simply because the resolution is far superior, provided that sample alignment is accurate and that all the corrections for the optical path through the crystal have been correctly made.

As an example of the value of detailed measurements of the complete set of phonon frequencies and their assignments, we cite the work on Mg_2SiO_4 (Price et al. 1987a,b). A set of models has recently been evaluated by comparing calculated phonon frequencies with the measured values. One model was shown to be far better than the others tested; this model has since been widely used as a transferable model for silicates (Winkler et al. 1991a).

Measurements of soft modes

Vibrational spectroscopy is particularly suited for the measurements of soft modes. Historically Raman spectroscopy has been the more useful technique, for until recently infrared absorption methods could not work at sufficiently low frequencies. Some of the earliest studies of soft modes were on quartz (Landsberg and Mandelstam 1929; Raman and Nedungadi 1940; Shapiro et al. 1967; Scott 1968; Berge et al. 1986; see also the review by Dolino 1990), but now a large number of soft mode phase transitions have been studied by spectroscopy (Scott 1974). Apart from simply measuring the temperature dependence of the soft mode frequency, which itself was a valuable contribution, many studies were also able to provide information on the behaviour of the linewidth of the soft mode on approaching the transition temperature. It was often found that the soft mode becomes heavily damped or even overdamped on approaching the transition temperature. One problem in this type of work is that the selection rules may limit spectroscopic observations to the low-temperature phase only. For example, the soft mode in $SrTiO_3$ has the wave vector at the Brillouin zone boundary of the high-temperature phase, so it can be measured only in the low-temperature phase where the zone boundary point becomes a reciprocal lattice vector (Fleury et al. 1968), and in many cases the selection rules for the soft mode at $\mathbf{k} = 0$ mean that it can be observed by spectroscopy only in the low-symmetry phase.

Measurements of the effects of phase transitions on the phonon spectra: hard mode spectroscopy

The frequencies, linewidths and intensities of the peaks in the vibrational spectra will be sensitive functions of the detailed structural state of the crystal. This feature can be exploited for quantitative studies of phase transitions, including displacive and order–disorder transitions, and electronic and magnetic ordering transitions. We have already described in Chapter 8 how the frequency of a mode will vary with the size of the order parameter, enabling the temperature dependence of the order parameter to be obtained from careful measurements of peak frequencies. A similar behaviour will also occur for order–disorder phase transitions. Structural changes will affect the intensities of spectral peaks, especially when a particular mode can be observed only in one phase due to the operation of the selection rules in the higher-symmetry phase. Furthermore, the linewidths will also be sensitive to structural changes. The realisation that there is considerable information that can be extracted from the high-frequency modes has resulted in many recent studies of these modes. The phrase *hard mode spectroscopy* has been coined for this area of application (Petzelt and Dvorak 1976a,b, 1984; Bismayer 1988; Salje 1992).

We must remark on the fact that the high-frequency modes will have frequencies that are virtually independent of wave vector. This means that the bonds vibrate almost independently, so that spectroscopic measurements are in fact acting as a local probe rather than a probe of collective motion[14] (Salje 1992). For example, two Si–O bonds that have different environments in a disordered structure will vibrate with slightly different frequencies, both of which may be recorded in one vibrational spectrum. Moreover, in disordered high-temperature phases, the local structure may be significantly different from the average structure. This can occur when the high-temperature phase consists of small domains or clusters that have the local structure of the low-temperature ordered phase, but when the structure is averaged over all the domains the net structure corresponds more closely to that of the higher-symmetry phase. In this case the frequencies of the high-energy modes may be more sensitive to the local order than the average structure, and can therefore give information about the local ordering. One example of this is the mineral cordierite, $Mg_2Si_5Al_4O_{18}$, which has a high-temperature phase in which the Al and Si atoms are disordered over tetrahedral sites. When crystalline cordierite is annealed from a glass of the correct stoichiometry, the first phase formed is the disordered phase. Further annealing leads to the transformation to the ordered form via a well-defined incommensurate intermediate phase (Putnis et al. 1987). The infrared absorption (Güttler et al. 1989) and Raman (Poon et al. 1990) spectra of cordierite show peaks with frequencies that are apparently insensitive to the intermediate incommensurate phase, which has led to the suggestion that the local structures of the ordered and incommensurate phases are identical.

The clearest use of measurements of peak intensities is when the intensity of a mode vanishes in the high-temperature phase, in which the mode is inactive. In this case, the intensity gives direct information on the magnitude of the order parameter, in a method analogous to the use of the measurements of the

[14] This argument should not be applied to low-frequency modes, for which there is a significant dependence of the phonon frequency on wave vector. In the language of Chapter 3, we are using the approximation of very different force constants, equation (3.26).

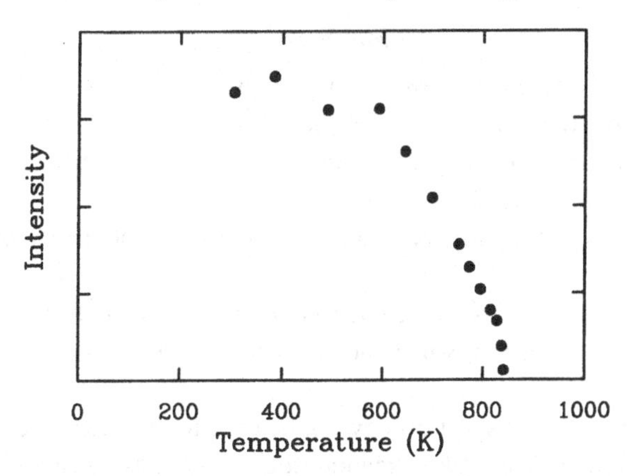

Figure 10.3: Temperature dependence of the intensity of a vibrational mode of quartz that is absent in the high-temperature phase (Salje et al. 1991). There is a first-order phase transition at 847 K. The temperature dependence of the intensity follows the square of the order parameter.

intensities of superlattice X-ray or neutron diffraction peaks, and complementary to the use of changes in the peak frequencies. We have remarked above that high-frequency modes are sensitive to local order. There may be cases when a high-frequency mode should be required to be absent by symmetry in the high-temperature phase but in practice it does not actually vanish on heating into the high-temperature phase. This observation can be accounted for by invoking the existence of domains that have the local structure of the low-temperature phase. Thus spectroscopy can be used as a sensitive test of models of phase transitions that postulate clusters in the high-temperature phases. For example, recent studies of the orientational order–disorder phase transition in $NaNO_3$ have shown the existence of a mode that is required to be absent in the high-temperature phase but which remains relatively strong on heating above the transition temperature (Harris et al. 1990). This observation points to the existence of considerable short-range order in the disordered phase. An opposite set of examples concerns quartz (Salje et al. 1992) and cristobalite (Swainson and Dove, 1992), for which spectroscopic measurements have shown that, contrary to the interpretation of diffraction data, the high-temperature phases do not consist of domains of the low-temperature phase. Experimental infrared absorption data for a single mode in quartz are shown in Figure 10.3; the data show unambiguously the vanishing of the intensity in the high-temperature phase.

Summary

1 Infrared and Raman spectroscopic techniques measure phonons with wave vectors close to the centre of the Brillouin zone.
2 Infrared absorption occurs for normal modes that change the dipole moment of the unit cell.
3 Raman spectroscopy measures the normal modes that change the polarisability.
4 Infrared and Raman spectroscopic techniques have a number of advantages over neutron scattering, which include better resolution and higher intensities.
5 Infrared and Raman spectroscopy can be used for a wide range of applications, including phase identification, local structures, accurate frequency measurements, measurements of soft modes as functions of temperature or pressure, and studies of phase transitions via the effects of the transitions on high-frequency modes.

FURTHER READING

Blinc and Zeks (1974)
Brüesch (1986) ch. 2–4
Iqbal and Owens (1984)

11

Formal quantum-mechanical description of lattice vibrations

The quantum-mechanical version of lattice dynamics is developed. We introduce the important creation and annihilation operators, and the Hamiltonian is recast in this new formalism. A number of results are obtained using the quantum formalism.

Some preliminaries

We have previously introduced the quantum-mechanical nature of lattice vibrations by noting the analogy with electromagnetic vibrations (phonons instead of photons) and simply modifying the classical model accordingly. On the other hand, the phonon picture emerges quite naturally when we start from a quantum-mechanical basis. It is the purpose of this chapter to introduce the essential ideas of the quantum-mechanical picture of lattice vibrations. You may recall from the preface that this is a chapter that can be ignored by the worried reader without any loss to the rest of the book. However, many professional solid state physicists like to start their theoretical discussions using the quantum-mechanical basis (e.g. Bruce and Cowley 1981, p 6), with the result that the reader who is unfamiliar with the basic concepts is effectively excluded from large sections of the research literature.

Most of us meet quantum mechanics in the form of the Schrödinger equation and the wave function, ψ (e.g. Rae 1981). The one-dimensional Schrödinger equation in its simple form for a single particle of mass m with associated position coordinate q and momentum p (= $m\dot{q}$) is given by

$$\mathcal{H}(p,q)\psi(p,q) = E\psi(p,q) \tag{11.1}$$

where $\mathcal{H}(p, q)$ is the Hamiltonian, $\psi(p, q)$ is the wave function, and E is the corresponding energy. The Schrödinger equation is an eigenequation, with ψ as the eigenvector and E as the eigenvalue. In the formalism of quantum mechanics, $\mathcal{H}$, q and p enter equation (11.1) as operators $\hat{\mathcal{H}}$, $\hat{q}$ and $\hat{p}$, such that

$$\hat{q}\psi = q\psi \tag{11.2a}$$

$$\hat{p}\psi = -i\hbar\frac{\partial}{\partial q}\psi \tag{11.2b}$$

If the particle experiences a potential V that depends only on q, the Hamiltonian operator of equation (11.1) can be written as

$$\hat{\mathcal{H}} = \frac{\hat{p}^2}{2m} + V(\hat{q}) = -\frac{\hbar^2}{2m}\frac{\partial^2}{\partial q^2} + V(\hat{q}) \tag{11.3}$$

The average (or most probable) value of q (called the *expectation value* in the language of quantum mechanics) is obtained from the integral:

$$\langle q \rangle = \int \psi^* \hat{q} \psi \, \mathrm{d}q \tag{11.4}$$

where * denotes the complex conjugate. Replacing $\hat{q}$ by $\hat{p}$ in equation (11.4) will give the most probable value for the momentum.

The wave function ψ_j for a single particle state is often written as $|j\rangle$, where j is the label of a given excited state (i.e. j labels one solution of equation (11.1)); ψ_j^* is then written as $\langle j|$. In the same language equation (11.4) is written as

$$\langle q \rangle = \langle j|\hat{q}|j\rangle \tag{11.5}$$

As well as describing the energy levels of a system, the wave function can also describe systems with more than one particle. If n_j gives the number of particles in the j-th state of a system, the wave function can be written as $|n_1, n_2, n_3, \ldots, n_j, \ldots\rangle$. For particles subject to the Pauli exclusion principle (called *fermions*), such as electrons, the values of n_j are restricted to zero or one. The quanta of lattice vibrations (phonons) are, however, not subject to this restriction, so that n_j can take on any value, including zero (the family of particles with this property are called *bosons* – any particle is either a fermion or a boson). The case of $n_j = 0$ for all states j is the lowest energy state, and is known as the *ground state* (it is the state for the temperature of absolute zero), and in this case the wave function is simply written as $|0\rangle$. In the general case where the particular state of a wave function is not specified, ψ is simply written as $|\,\rangle$.

One important concept in the formalism of quantum mechanics is that of commutation. For two operators, say $\hat{q}$ and $\hat{p}$, we define the *commutation relation* as

$$[\hat{q}, \hat{p}] = \hat{q}\hat{p} - \hat{p}\hat{q} \tag{11.6}$$

In this specific case:

$$[\hat{q}, \hat{p}] = i\hbar \tag{11.7}$$

If the commutation relation equals zero, we say that the two operators *commute.* The concept of commutation is used in many applications of quantum mechanics, as we will see below. It turns out that two operators that have a Heisenberg uncertainty relation will not commute. All the important commutation relations that are used in this chapter are given in Appendix G.

Quantum-mechanical description of the harmonic crystal

The Hamiltonian for the harmonic crystal in terms of the normal mode coordinates was described in detail in Chapters 4 and 6 (equations (4.7) and (6.44)). In its operator form, it can be written as

$$\hat{\mathcal{H}} = \frac{1}{2}\sum_{\mathbf{k},\nu}\left\{\hat{P}(\mathbf{k},\nu)\hat{P}^{+}(\mathbf{k},\nu) + \omega^2(\mathbf{k},\nu)\hat{Q}(\mathbf{k},\nu)\hat{Q}^{+}(\mathbf{k},\nu)\right\} \tag{11.8}$$

where $\hat{Q}(\mathbf{k}, \nu)$ is the operator corresponding to the normal mode coordinate, and $\hat{P}(\mathbf{k}, \nu)$ is the corresponding momentum variable. The superscript $^+$ denotes the adjoint of the operators. We note that unlike the real space operators the normal mode coordinate operators are not Hermitian, i.e.

$$\left.\begin{array}{l}\hat{q}_j^{+} = \hat{q}_j \\ \hat{p}_j^{+} = \hat{p}_j\end{array}\right\} \Rightarrow \begin{array}{l}\hat{Q}^{+}(\mathbf{k},\nu) = \hat{Q}(-\mathbf{k},\nu) \\ \hat{P}^{+}(\mathbf{k},\nu) = \hat{P}(-\mathbf{k},\nu)\end{array} \tag{11.9}$$

It turns out that the picture looks a lot simpler if we define two new operators $\hat{a}(\mathbf{k}, \nu)$ and $\hat{a}^{+}(\mathbf{k}, \nu)$:[1]

$$\hat{a}(\mathbf{k},\nu) = \frac{1}{\left(2\hbar\omega(\mathbf{k},\nu)\right)^{1/2}}\left\{\omega(\mathbf{k},\nu)\hat{Q}(\mathbf{k},\nu) + i\hat{P}(\mathbf{k},\nu)\right\} \tag{11.10a}$$

[1] In some books (e.g. Inkson, 1984, pp 46–47) the following definitions are rotated in the complex plane. In discussions that are limited to the monatomic chain the atomic mass is sometimes also included explicitly in the following definitions (also in Inkson 1984, pp 46–47). Neither of these variants leads to different conclusions.

$$\hat{a}^+(\mathbf{k},\nu)=\frac{1}{\left(2\hbar\omega(\mathbf{k},\nu)\right)^{1/2}}\left\{\omega(\mathbf{k},\nu)\hat{Q}^+(\mathbf{k},\nu)-i\hat{P}^+(\mathbf{k},\nu)\right\} \quad (11.10b)$$

According to these definitions, the operators for the normal mode coordinates can be written as

$$\hat{Q}(\mathbf{k},\nu)=\left(\frac{\hbar}{2\omega(\mathbf{k},\nu)}\right)^{1/2}\left(\hat{a}(\mathbf{k},\nu)+\hat{a}^+(-\mathbf{k},\nu)\right) \quad (11.11a)$$

$$\hat{P}(\mathbf{k},\nu)=-i\left(\frac{\hbar\omega(\mathbf{k},\nu)}{2}\right)^{1/2}\left(\hat{a}(\mathbf{k},\nu)-\hat{a}^+(-\mathbf{k},\nu)\right) \quad (11.11b)$$

The corresponding real space operators (from equation (4.2)) are similarly related to our new operators by

$$\hat{\mathbf{q}}(jl)=\frac{1}{\left(Nm_j\right)^{1/2}}\sum_{\mathbf{k},\nu}\left\{\left(\frac{\hbar}{2\omega(\mathbf{k},\nu)}\right)^{1/2}\mathbf{e}(j,\mathbf{k},\nu)\exp\left(i\mathbf{k}\cdot\mathbf{r}(jl)\right)\right.$$
$$\left.\times\left(\hat{a}(\mathbf{k},\nu)+\hat{a}^+(-\mathbf{k},\nu)\right)\right\} \quad (11.12a)$$

$$\hat{\mathbf{p}}(jl)=\frac{1}{\left(Nm_j\right)^{1/2}}\sum_{\mathbf{k},\nu}\left\{\left(\frac{\hbar\omega(\mathbf{k},\nu)}{2}\right)^{1/2}\mathbf{e}(j,\mathbf{k},\nu)\exp\left(i\mathbf{k}\cdot\mathbf{r}(jl)\right)\right.$$
$$\left.\times\left(\hat{a}(\mathbf{k},\nu)-\hat{a}^+(-\mathbf{k},\nu)\right)\right\} \quad (11.12b)$$

When we substitute equation (11.11) into equation (11.8), noting the relations (11.9), we obtain a new form for the Hamiltonian

$$\hat{\mathcal{H}}=\frac{1}{4}\sum_{\mathbf{k},\nu}\hbar\omega(\mathbf{k},\nu)\left\{\left(\hat{a}(\mathbf{k},\nu)-\hat{a}^+(-\mathbf{k},\nu)\right)\left(\hat{a}^+(\mathbf{k},\nu)-\hat{a}(-\mathbf{k},\nu)\right)\right.$$
$$\left.+\left(\hat{a}(\mathbf{k},\nu)+\hat{a}^+(-\mathbf{k},\nu)\right)\left(\hat{a}^+(\mathbf{k},\nu)+\hat{a}(-\mathbf{k},\nu)\right)\right\}$$
$$=\frac{1}{2}\sum_{\mathbf{k},\nu}\hbar\omega(\mathbf{k},\nu)\left\{\hat{a}(\mathbf{k},\nu)\hat{a}^+(\mathbf{k},\nu)+\hat{a}^+(\mathbf{k},\nu)\hat{a}(\mathbf{k},\nu)\right\}$$
$$=\frac{1}{2}\sum_{\mathbf{k},\nu}\hbar\omega(\mathbf{k},\nu)\left\{\left[\hat{a}(\mathbf{k},\nu),\hat{a}^+(\mathbf{k},\nu)\right]+2\hat{a}^+(\mathbf{k},\nu)\hat{a}(\mathbf{k},\nu)\right\}$$
$$=\sum_{\mathbf{k},\nu}\hbar\omega(\mathbf{k},\nu)\left(\hat{a}^+(\mathbf{k},\nu)\hat{a}(\mathbf{k},\nu)+\frac{1}{2}\right) \quad (11.13)$$

This looks a lot simpler than equation (11.8)! In deriving equation (11.13) we have used the commutation relation (G.10).

The new operators: creation and annihilation operators

Before we can exploit the form of our new Hamiltonian, we need to give some meaning to the operators $\hat{a}(\mathbf{k}, \nu)$ and $\hat{a}^+(\mathbf{k}, \nu)$. To make the nomenclature easier, we denote ($\mathbf{k}$, ν) by k and ($-\mathbf{k}$, ν) by $-k$. Consider a wave function $|\psi\rangle$ that is an eigenvector of $\hat{\mathcal{H}}$, such that

$$\hat{\mathcal{H}}|\psi\rangle = \sum_k \hbar\omega_k\left(\hat{a}_k^+\hat{a}_k + \frac{1}{2}\right)|\psi\rangle = E|\psi\rangle \tag{11.14}$$

Now consider the new wave function $\hat{a}_k^+|\psi\rangle$. We then have:

$$\hat{\mathcal{H}}\hat{a}_k^+|\psi\rangle = \hbar\omega_k\left(\hat{a}_k^+\hat{a}_k\hat{a}_k^+ + \frac{1}{2}\hat{a}_k^+\right)|\psi\rangle \tag{11.15}$$

Using the commutation relation (G.11a) we obtain the result:

$$\begin{aligned}\hat{\mathcal{H}}\hat{a}_k^+|\psi\rangle &= \left(\hbar\omega_k\hat{a}_k^+ + \hat{a}_k^+\hat{\mathcal{H}}\right)|\psi\rangle \\ &= (E + \hbar\omega_k)\hat{a}_k^+|\psi\rangle \end{aligned} \tag{11.16a}$$

Similarly, using the commutation relation (G.11b), we can show that

$$\hat{\mathcal{H}}\hat{a}_k|\psi\rangle = (E - \hbar\omega_k)\hat{a}_k|\psi\rangle \tag{11.16b}$$

We have therefore shown that both $\hat{a}_k^+|\psi\rangle$ and $\hat{a}_k|\psi\rangle$ are eigenvectors of $\hat{\mathcal{H}}$, with the respective energies $E + \hbar\omega_k$ and $E - \hbar\omega_k$. It thus appears that the effect of the operator $\hat{a}_k^+$ is to increase the energy by one quantum of $\hbar\omega_k$, whereas the effect of the operator $\hat{a}_k$ is to decrease the energy by one quantum of $\hbar\omega_k$. For this reason $\hat{a}_k^+$ and $\hat{a}_k$ are called *creation* and *annihilation* operators respectively, since they create or annihilate quanta of lattice vibrations (phonons). Alternative names are *raising* and *lowering* operators, or *ladder* operators.

The Hamiltonian and wave function with creation and annihilation operators

The ground state energy

Let us start with the ground state wave function $|0\rangle$. The definition of the ground state means that:

$$\hat{a}_k|0\rangle = 0 \tag{11.17}$$

Hence we have:

$$\begin{aligned}\hat{\mathcal{H}}|0\rangle &= \sum_k \hbar\omega_k\left(\hat{a}_k^+\hat{a}_k + \frac{1}{2}\right)|0\rangle \\ &= \sum_k \frac{1}{2}\hbar\omega_k|0\rangle\end{aligned} \tag{11.18}$$

This means that the energy of the ground state is not zero but is equal to

$$E_0 = \sum_k \frac{1}{2}\hbar\omega_k \tag{11.19}$$

This energy is called the *zero-point energy*. The existence of the zero-point energy, which we encountered in Chapter 4, implies that the atoms in the crystal can never be at rest even at a temperature of absolute zero.[2]

Normalisation of the wave function

Consider now the wave function for the state k containing n_k phonons, $|n_k\rangle$, defined with an amplitude α which we have not yet determined:

$$|n_k\rangle = \alpha\left(\hat{a}_k^+\right)^{n_k}|0\rangle \tag{11.20}$$

To obtain the normalisation of $|n_k\rangle$ we consider the following operations:

[2] Zero-point motion is particularly important for helium, as it prevents freezing of the liquid at low pressure.

$$\begin{aligned}\hat{a}_k^+\hat{a}_k|n_k\rangle &= \alpha\hat{a}_k^+\hat{a}_k\left(\hat{a}_k^+\right)^{n_k}|0\rangle \\ &= \alpha\hat{a}_k^+\left(\hat{a}_k\hat{a}_k^+\right)\left(\hat{a}_k^+\right)^{n_k-1}|0\rangle \\ &= \alpha\hat{a}_k^+\left(\hat{a}_k^+\hat{a}_k+1\right)\left(\hat{a}_k^+\right)^{n_k-1}|0\rangle \\ &= \alpha\left(\hat{a}_k^+\right)^2\hat{a}_k\left(\hat{a}_k^+\right)^{n-1}|0\rangle+|n_k\rangle\end{aligned} \tag{11.21}$$

We repeat the same procedure for the first term in the last line of equation (11.21) until we obtain the end result:

$$\hat{a}_k^+\hat{a}_k|n_k\rangle = \alpha\left(\hat{a}_k^+\right)^{n_k+1}\hat{a}_k|0\rangle + n_k|n_k\rangle = n_k|n_k\rangle \tag{11.22}$$

Equation (11.22) therefore defines the *number operator*, $\hat{a}_k^+\hat{a}_k$, which gives the number of phonons as the eigenvalue of the wave function. Thus the phonon Hamiltonian, equation (11.13), has the eigenvalues:

$$\hat{\mathcal{H}}|n_k\rangle = \hbar\omega_k\left(n_k+\frac{1}{2}\right)|n_k\rangle \tag{11.23}$$

Using the commutation relation (G.10a) we also obtain the result analogous to equation (11.22):

$$\hat{a}_k\hat{a}_k^+|n_k\rangle = (n_k+1)|n_k\rangle \tag{11.24}$$

The wave function is normalised such that:

$$\langle n_k|n_k\rangle = 1 \tag{11.25}$$

Hence,

$$\langle n_k|\hat{a}_k\hat{a}_k^+|n_k\rangle = n_k+1 \tag{11.26a}$$

$$\langle n_k|\hat{a}_k^+\hat{a}_k|n_k\rangle = n_k \tag{11.26b}$$

These results suggest the following normalisations:

$$\hat{a}_k^+|n_k\rangle = (n_k+1)^{1/2}|n_k+1\rangle \tag{11.27a}$$

$$\hat{a}_k|n_k\rangle = n_k^{1/2}|n_k-1\rangle \tag{11.27b}$$

From equation (11.27a) we have the following identity:

$$\left(\hat{a}_k^+\right)^{n_k}|0\rangle = \left(n_k!\right)^{1/2}|n_k\rangle \tag{11.28}$$

which leads to the normalisation of the wave function:

$$|n_k\rangle = \frac{\left(\hat{a}_k^+\right)^{n_k}}{\left(n_k!\right)^{1/2}}|0\rangle \tag{11.29}$$

thereby defining the value of α in equation (11.19)

Time and position dependence

The formalism using creation and annihilation operators gives information concerning the phonon interactions and amplitudes, but does not contain time or position dependence. Thus we know, for example, how two phonons merge to form a third, but we do not know the time dependence of this process. We can define the following time-dependent operators:

$$\hat{a}_k(t) = \hat{a}_k \exp(-i\omega_k t) \tag{11.30a}$$

$$\hat{a}_k^+(t) = \hat{a}_k^+ \exp(i\omega_k t) \tag{11.30b}$$

To include the spatial dependence as well, we define the following operators:

$$\hat{\psi}(\mathbf{r},t) = \sum_k \phi_k(\mathbf{r})\hat{a}_k(t) \tag{11.31a}$$

$$\hat{\psi}^+(\mathbf{r},t) = \sum_k \phi_k^+(\mathbf{r})\hat{a}_k^+(t) \tag{11.31b}$$

where $\phi_k(\mathbf{r})$ is the wave function for the state k. These new operators are called *field operators*, and a detailed analysis shows that despite the fact that they are operators, they behave like wave functions and obey the Schrödinger equation. However, unlike the Schrödinger equation, the form of quantum mechanics with the field operators also contains information about the statistics of the system. The field operators obey the following commutation relation:

$$\begin{aligned}\left[\hat{\psi}(\mathbf{r},t), \hat{\psi}^+(\mathbf{r}',t)\right] &= \sum_{k,k'} \phi_k(\mathbf{r})\phi_{k'}^+(\mathbf{r}')\left[\hat{a}_k(t), \hat{a}_{k'}^+(t)\right] \\ &= \sum_k \phi_k(\mathbf{r})\phi_k^+(\mathbf{r}') \\ &= \delta(\mathbf{r}-\mathbf{r}')\end{aligned} \tag{11.32}$$

At this point we have reached about as far in the development of the quantum theory as we can usefully go. The formalism at this stage is generally given the name *second quantisation*. In the form described here a number of results used in this book can be derived (see below). Perhaps of greater importance is the fact that the second quantisation formalism permits a straightforward method of incorporating interactions between different types of excitation, one of the most important of which is the electron–phonon interaction. The development of quantum mechanics described in this chapter has been appropriate for bosons only, but analogous results can be obtained for fermions.

Applications

Normal mode amplitude

From the definition of the normal mode coordinate, equations (11.9)–(11.11), we note the operator product:

$$\hat{Q}_k\hat{Q}_k^+ = \frac{\hbar}{2\omega_k}\left(\hat{a}_k + \hat{a}_{-k}^+\right)\left(\hat{a}_{-k} + \hat{a}_k^+\right)$$

$$= \frac{\hbar}{2\omega_k}\left(\hat{a}_k\hat{a}_{-k} + \hat{a}_{-k}^+\hat{a}_k^+ + \hat{a}_k\hat{a}_k^+ + \hat{a}_{-k}^+\hat{a}_{-k}\right) \tag{11.33}$$

The terms that couple different wave functions vanish:

$$\langle \ |\hat{a}_k\hat{a}_{-k}|\ \rangle = \langle \ |\hat{a}_{-k}^+\hat{a}_k^+|\ \rangle = 0 \tag{11.34}$$

The remaining terms are given by equation (11.26), leading to the result for the expectation value of $\hat{Q}_k\hat{Q}_k^+$:

$$\langle n_k|\hat{Q}_k\hat{Q}_k^+|n_k\rangle = \frac{\hbar}{2\omega_k}\left(\langle n_k|\hat{a}_k\hat{a}_k^+|n_k\rangle + \langle n_k|\hat{a}_{-k}^+\hat{a}_{-k}|n_k\rangle\right)$$

$$= \frac{\hbar}{2\omega_k}(2n_k+1)\langle n_k|n_k\rangle$$

$$= \frac{\hbar}{2\omega_k}(2n_k+1) \tag{11.35}$$

We obtained the same result in equation (4.15).

One-phonon scattering cross section and detailed balance

The important correlation function for one-phonon neutron scattering is

$$\left\langle \left[\mathbf{Q}\cdot\mathbf{u}_j(t)\right]\left[\mathbf{Q}\cdot\mathbf{u}_{j'}(0)\right]\right\rangle \tag{11.36}$$

where $\mathbf{Q}$ is the scattering vector. Using equation (11.12a) we obtain:

$$\mathbf{Q}\cdot\mathbf{u}_j(t) = \left(\frac{\hbar}{2Nm_j}\right)^{1/2} \sum_{\mathbf{k},\nu} \frac{1}{\sqrt{\omega(\mathbf{k},\nu)}} \Big\{\mathbf{Q}\cdot\mathbf{e}(j,\mathbf{k},\nu)\exp\big(i\mathbf{k}\cdot\mathbf{r}(jl)\big)$$
$$\times\big(\hat{a}(\mathbf{k},\nu)\exp(-i\omega(\mathbf{k},\nu)t) + \hat{a}^+(-\mathbf{k},\nu)\exp(i\omega(\mathbf{k},\nu)t)\big)\Big\} \tag{11.37}$$

Therefore the correlation function (11.36) can be written as

$$\left\langle \left[\mathbf{Q}\cdot\mathbf{u}_j(t)\right]\left[\mathbf{Q}\cdot\mathbf{u}_{j'}(0)\right]\right\rangle = \frac{\hbar}{2N\sqrt{m_j m_{j'}}} \sum_{\mathbf{k},\nu}\sum_{\mathbf{k}',\nu'} \Bigg\{ \frac{1}{\sqrt{\omega(\mathbf{k},\nu)\omega(\mathbf{k}',\nu')}}$$
$$\times\left[\mathbf{Q}\cdot\mathbf{e}(j,\mathbf{k},\nu)\right]\left[\mathbf{Q}\cdot\mathbf{e}^*(j',\mathbf{k}',\nu')\right]\exp\big(i\mathbf{k}\cdot\mathbf{r}(jl)\big)\exp\big(i\mathbf{k}'\cdot\mathbf{r}(j'l')\big)$$
$$\times\big(\hat{a}(\mathbf{k},\nu)\exp(-i\omega(\mathbf{k},\nu)t) + \hat{a}^+(-\mathbf{k},\nu)\exp(i\omega(\mathbf{k},\nu)t)\big)$$
$$\times\big(\hat{a}(\mathbf{k}',\nu') + \hat{a}^+(-\mathbf{k}',\nu')\big)\Bigg\} \tag{11.38}$$

We note that the operator products will only be non-zero if $\mathbf{k}' = -\mathbf{k}$ and $\nu' = \nu$. We also note the expectation values for pairs of operators, equations (11.22), (11.24) and (11.34), to obtain the final result:

$$\left\langle \left[\mathbf{Q}\cdot\mathbf{u}_j(t)\right]\left[\mathbf{Q}\cdot\mathbf{u}_{j'}(0)\right]\right\rangle = \frac{\hbar}{2N\sqrt{m_j m_{j'}}} \sum_{\mathbf{k},\nu} \Bigg\{ \frac{1}{\omega(\mathbf{k},\nu)}$$
$$\times\left[\mathbf{Q}\cdot\mathbf{e}(j,\mathbf{k},\nu)\right]\left[\mathbf{Q}\cdot\mathbf{e}^*(j',-\mathbf{k},\nu)\right]\exp\big(i\mathbf{k}\cdot\left[\mathbf{r}(j)-\mathbf{r}(j')\right]\big)$$
$$\times\left[(n(\omega)+1)\exp(-i\omega(\mathbf{k},\nu)t) + n(\omega)\exp(i\omega(\mathbf{k},\nu)t)\right]\Bigg\} \tag{11.39}$$

Fourier transformation of equation (11.39) leads simply to the one-phonon neutron scattering function, equation (9.18). The use of creation and annihilation operators has served to produce one important aspect that would not follow from classical theory. The probability of a neutron being scattered following creation of a phonon is higher than the probability of scattering following absorption of a phonon energy by the ratio $(n + 1)/n$. The difference between the two processes is known as *detailed balance*, and arises because of the dependence of the scattering medium on its original state. For absorption of phonons to occur there already needs to be a population of excited phonons. The state of the scattering system cannot be taken into account properly in a classical treatment, so that the classical neutron scattering function will be the same for both energy gain and energy loss.[3]

Anharmonic interactions

We consider the cubic term in the expansion of the potential energy:

$$V^{(3)}_{k,k',k''}\hat{Q}_k\hat{Q}_{k'}\hat{Q}_{k''} = \frac{V^{(3)}_{k,k',k''}}{\sqrt{\omega_k\omega_{k'}\omega_{k''}}}\left(\frac{\hbar}{2}\right)^{3/2}\left(\hat{a}_k + \hat{a}^+_{-k}\right)\left(\hat{a}_{k'} + \hat{a}^+_{-k'}\right)\left(\hat{a}_{k''} + \hat{a}^+_{-k''}\right) \tag{11.40}$$

This term generates four different types of terms:

$\hat{a}_k\hat{a}_{k'}\hat{a}_{k''}$: Three phonons are spontaneously destroyed.

$\hat{a}^+_k\hat{a}^+_{k'}\hat{a}^+_{k''}$: Three phonons are spontaneously created.

$\hat{a}_k\hat{a}^+_{k'}\hat{a}^+_{k''}$: The phonon k is destroyed and the phonons k' and k'' are created. This corresponds to the decay of k into the two phonons k' and k''. There are two other similar terms.

$\hat{a}_k\hat{a}_{k'}\hat{a}^+_{k''}$: The phonons k and k' are destroyed and the phonon k'' is created. This corresponds to a collision of the two phonons k and k', which coalesce to form the phonon k''. There are two other similar terms.

The first two terms do not conserve energy or wave vector, but in principle can occur as virtual processes.[4] The last two terms can conserve energy and wave vector, and are therefore of considerable importance. Similar results follow for higher-order anharmonic terms: the n-th order anharmonic interactions involve processes with n phonons. The picture that emerges from the use of creation and annihilation operators tends to highlight our visualisation of phonons as actual particles.

[3] This point is made at the end of Appendix E.

[4] Virtual processes are of importance in quantum field theory, as they occur over time scales that are related to the violation of the conservation of energy by the corresponding uncertainty principle.

Summary

1 The quantum-mechanical operators for the normal mode coordinates can be expressed in terms of two new operators, called creation and annihilation operators. The names of these operators arise because their effects are to create and annihilate phonons respectively.
2 The phonon Hamiltonian can be recast in terms of the creation and annihilation operators.
3 The wave functions can also be recast in terms of the creation and annihilation operators.
4 The creation and annihilation operators can be combined to give information about the number of phonons that exist for a given state of the system.
5 A number of results have been obtained using the new formalism.

FURTHER READING

Ashcroft and Mermin (1976) app. L–O
Brüesch (1982) pp 33–45, 69–72; App E
Inkson (1984) ch. 3, 4
Rae (1981)

12

Molecular dynamics simulations

Molecular dynamics simulations provide a method for studying crystals at high temperatures, taking full account of anharmonic interactions. The essentials of the method are described, and technical aspects of the calculation of normal mode coordinates and phonon frequencies are discussed in some detail.

The molecular dynamics simulation method

We have noted that anharmonic effects are difficult to calculate in practice. Yet these are often of considerable interest, as for example at a phase transition or in a disordered system. The extreme example of the effect of anharmonicity is the melting phase transition, about which anharmonic phonon theory is practically unable to give very much quantitative information. One approach to these sorts of problem might be to say that theory can give a qualitative understanding at least, which we can then use to help interpret experimental data. We don't, after all, always need a quantitative theory to understand phenomena we can measure. But this approach will not always be enough, for it is often the case that qualitative theory cannot be matched with experimental results. It is for such cases that computer models have been developed. We will consider one computer simulation method, the *molecular dynamics simulation* (MDS) technique, which has proved to be particularly useful.

The essence of the MDS method is to solve Newton's equation of motion (equation (1.1)) for a set of fictitious atoms, whose coordinates and velocities are stored in a computer, and which are assumed to interact via a model interatomic anharmonic potential. Any numerical solution of the continuous equations of motion will involve using small discrete time steps, with a method (the algorithm) to generate the atomic positions and velocities at a given time step from the positions and velocities of the previous time steps. The MDS method will then be able to generate the *classical* trajectories of the collection of atoms over a period of time that is long enough to be able to analyse with adequate accuracy. The particular algorithm and the size of the time step will be chosen so that the numerical solution will not contain any significant errors, but also so that the simulation can be carried out within the constraints of available computer power.

The MDS method was invented by Alder and Wainwright (1957, 1959) and applied to a fluid of hard spheres. The motivation was to go beyond the limitations of the theory and experimental methods of the day in a search for new insights into the behaviour of fluids. Hard sphere models do not give a true liquid state, and so the next stage in the development of the MDS technique was to use realistic interatomic potentials. This was first implemented for a model of liquid argon using Lennard-Jones interactions (Rahman 1964, 1966). Further work established that the MDS method is capable of giving results that are in quantitative agreement with experimental data. Simulations were then also performed on ionic fluids (Woodcock 1971, 1972), molecular fluids such as N_2 (Cheung and Powles 1975, 1976) and F_2 (Singer et al. 1977), and water (Rahman and Stillinger 1971, 1974). Because of its importance in chemistry, biology and geology, water has continued to be studied by MDS, particularly as new potential models are developed. A detailed discussion of the application of the MDS technique to liquids is given by Allen and Tildesley (1987).

The application of the MDS method to solids came later, with studies of argon, potassium (Hansen and Klein 1976), sodium chloride (Jacucci et al. 1976) and nitrogen (Weis and Klein 1975; Klein and Weis 1977). These were aimed at testing calculations of anharmonic phonon frequencies and investigating the effects of orientational disorder. The significant development given by these studies was the development of methods to study the lattice dynamics.

The primary aim of this chapter is to enable the reader to understand the essential details of the MDS method, particularly as applied to solids, so that the scientific literature reporting MDS studies is more accessible. I have resisted the temptation to attempt to help the reader become an expert practitioner – the book of Allen and Tildesley (1987) does this task already. But it is hoped that this chapter will be of some help to anyone wanting to use the MDS technique for themselves. Further reviews of the MDS method as applied to the study of solids are given by Klein (1978) and Dove (1988)[1].

[1] Which unfortunately contains more typographical errors than one would like.

Details of the molecular dynamics simulation method

Simulation algorithms

In its traditional formalism, the MDS method uses a constant number of atoms (usually a few hundreds or thousands) in an imaginary box which has a constant size and shape and periodic boundary conditions. All MDS studies use model interatomic potentials similar to those described in Chapter 1. As well as giving reasonable values for the first and second differentials, these models also give reasonable approximations of the anharmonic forces. At each time step, the forces between all atoms are calculated from these model potentials.

Newton's equation of motion is a continuous differential equation, which for computational studies needs to be approximated by a discrete integral equation. We can express the position of the j-th atom at a time $t+\Delta t$, $\mathbf{r}_j(t+\Delta t)$, as a Taylor expansion:

$$\mathbf{r}_j(t+\Delta t) = \mathbf{r}_j(t) + \sum_{n\geq 1} \frac{(\Delta t)^n}{n!} \frac{\partial^n \mathbf{r}_j(t)}{\partial t^n} \tag{12.1}$$

If we add to this the corresponding equation for $\mathbf{r}_j(t-\Delta t)$ and neglect the terms of order $(\Delta t)^4$ and higher, we can obtain a simple expression for the calculation of the particle position after a single time step:

$$\mathbf{r}_j(t+\Delta t) = 2\mathbf{r}_j(t) - \mathbf{r}_j(t-\Delta t) + \Delta t^2 \mathbf{F}_j(t) / m_j \tag{12.2}$$

where m_j is the mass of the j-th atom, and $\mathbf{F}_j(t)$ is the force experienced by this atom at time t. If on the other hand we subtract the equation for $\mathbf{r}_j(t-\Delta t)$ from the equation for $\mathbf{r}_j(t+\Delta t)$, we obtain an expression for the velocity of the atom at time t:

$$\dot{\mathbf{r}}_j(t) = \frac{1}{2\Delta t}\left(\mathbf{r}_j(t+\Delta t) - \mathbf{r}_j(t-\Delta t)\right) \tag{12.3}$$

These equations are known as the *Verlet algorithm*, after their inventor (Verlet 1967). The error in the expression for the velocity is of the order of $(\Delta t)^3$, whereas the error in the expression for the position is of the order $(\Delta t)^4$. These differences in accuracy are in fact unimportant, since the algorithm does not use the velocity in the calculation of the position at the next time step. Although this algorithm might appear to be a rather simple way of integrating the equations of motion, it turns out to give solutions that are sufficiently stable and accurate for routine use. In fact, it is hard to make major improvements to the algorithm without incurring the cost of increased computational effort. A number of simple variations of the Verlet algorithm are in common use (e.g. Beeman 1976).

It is quite common to simulate a system with periodic boundary conditions, in order that surface effects are removed.[2] The sample can be defined by three vectors, **X**, **Y** and **Z**, which are usually taken as multiples of the unit cell vectors. By comparison with the Born–von Kármán model, the number of wave vectors given by such a sample is limited to integral multiples of the three fundamental wave vectors **X***, **Y*** and **Z***:

$$\mathbf{X}^* = (\mathbf{Y} \times \mathbf{Z}) / V_s \tag{12.4a}$$

$$\mathbf{Y}^* = (\mathbf{Z} \times \mathbf{X}) / V_s \tag{12.4b}$$

$$\mathbf{Z}^* = (\mathbf{X} \times \mathbf{Y}) / V_s \tag{12.4c}$$

where V_s is the sample volume. If small samples (for example 64 unit cells) are used, the wave vector resolution is rather poor. Thus in the study of lattice dynamics by MDS, it is desirable to use as large a sample as possible. The use of 1000–4096 unit cells is a good compromise between the need for good wave vector resolution and the limitations of computing resources.

Thermodynamic ensembles

The method that has been described here is appropriate for the study of constant volume samples, and the equations of motion ensure that the *total* energy (potential energy + kinetic energy) remains constant throughout the simulation. The ensemble in which the number of particles, sample volume and total energy are conserved (constant) quantities is known as the *microcanonical ensemble*, and corresponds to an isolated system. This ensemble is particularly good for the study of dynamic processes (such as phonon propagation). But if we are interested in phase transitions, we may prefer to be able to let the sample size and shape relax in response to changes in temperature or pressure. For example, the cubic–tetragonal phase transition in cristobalite (SiO_2) leads to a 5% volume reduction on cooling through the transition (Schmahl et al. 1992). For such cases, Parrinello and Rahman (1980, 1981) invented a *constant pressure* algorithm in which the components of the three sample vectors **X**, **Y** and **Z**, are also treated as dynamical variables. The equations of motion

[2] When long-range interactions are included in the model, the use of periodic boundary conditions means that the Ewald method (Appendix A) can be used to evaluate these interactions. In this case the repeating unit is not the unit cell but the whole sample.

for the sample vector components are solved in exactly the same way as the equations for the positions, so that a new sample shape is generated at each time step. In this ensemble the conserved quantities are the number of particles, the pressure and the enthalpy.[3] A detailed review of the method, containing an extension for molecular crystals and a description of the application of the method to simulations using the Ewald method for handling long-range interactions, is given by Nosé and Klein (1983).

The standard experimental situation is that the sample is in thermal equilibrium with the environment; it is said that the sample is in contact with a heat bath of constant temperature. The ensemble in which the temperature of the heat bath is constant, and the number of particles and sample volume are conserved quantities, is called the *canonical ensemble*. This is the ensemble for which most theoretical work is appropriate. Nosé (1984a,b) has developed an algorithm for generating trajectories that are consistent with the canonical ensemble. One advantage of this method is that the sample temperature can be carefully controlled.[4] However, the main drawback of the method is that it gives an ambiguity concerning the dynamic properties, since it introduces a dynamic parameter that effectively scales the size of the time step. Nosé's method can be used for both constant volume and constant pressure simulations.

We have noted that the normal MDS equations of motion generate classical trajectories. For many properties quantum-mechanical effects are unimportant – for example, we have already demonstrated in this book that the values of the harmonic phonon frequencies are independent of whether they are obtained within a classical or quantum-theoretical framework – but quantum effects are certainly manifest in some quantities, even at room temperature. The clearest evidence of the importance of quantum effects is the heat capacity, which for many common materials has a value at room temperature that is significantly less than the classical Dulong–Petit value. A number of methods have been developed to incorporate quantum mechanics into the MDS method, which vary depending upon the motivation (Allen and Tildesley 1987; Matsui 1989). We will also see below that an alternative approach is to incorporate quantum mechanics into the analysis stage of a classical simulation rather than into the simulation method itself.

[3] In practice the enthalpy is not exactly conserved, but the fluctuations in the enthalpy are much smaller than the fluctuations in the kinetic or potential energies. The conserved quantity is the constant pressure Hamiltonian.

[4] The standard alternative approach to setting a desired temperature is to rescale the sample velocities at each time step until the simulation is equilibrated at a temperature close to the required temperature.

Sample size and time steps

The scope of systems that can be studied by MDS depends on the computing resources available. With smaller computers, it may only be possible to study samples with 64 or 125 unit cells, each containing only a small number of atoms. Larger computers, however, allow us to use samples of up to about 4096 unit cells with more than 20 atoms per unit cell (Tautz et al. 1991; Winkler and Dove 1992).

The size of the time step will depend upon the forces used in the simulation. The time step should certainly not be larger than about 5% of the shortest vibrational period – typical time steps will vary from 10^{-15} to 10^{-14} s. The use of a numerical algorithm for generating particle trajectories will necessarily introduce some error into the trajectories. Thus conserved quantities like the total energy in a simulation of a microcanonical ensemble will in practice fluctuate owing to these errors; the size of the effect will be related to the size of the time step. The time step is generally chosen as a compromise between accuracy and computational resources. A simulation should then be run for enough time steps so that the total simulated time is much longer than the longest period oscillation in the sample. For accurate calculations of single particle averages a few picoseconds will be adequate, but for calculations of collective properties some tens of picoseconds will be required.

The simulation can be separated into two stages. The first stage will be for equilibration of the sample, which will be completed once there is no long-time drift of any variable quantities, such as the temperature. Moreover, it is also important that the fluctuations settle into a pattern of constant amplitude. Only then can the second stage begin, which is the stage that generates reliable trajectories for detailed analysis.

Analysis of the results of a simulation

Macroscopic thermodynamics

Molecular dynamics simulations can give a range of thermodynamic information. The temperature T can be obtained from the atomic velocities in the normal classical manner, and the potential energy can be calculated from the model at each time step. The pressure P can be obtained from the standard *virial* expression:

$$P = \frac{Nk_{\mathrm{B}}T}{V_{\mathrm{s}}} - \frac{1}{3V_{\mathrm{s}}}\sum_j \mathbf{F}_j \cdot \mathbf{r}_j - \frac{\partial \varphi'}{\partial V_{\mathrm{s}}} \tag{12.5}$$

where $\mathbf{F}_j$ is the force experienced by the j-th particle at position $\mathbf{r}_j$, φ' is a contribution to the lattice energy that does not give rise to a force,[5] and N is the number of particles. Similarly, the stress tensor is given by

$$\sigma_{ij} = \frac{Nk_{\mathrm{B}}T}{V_{\mathrm{s}}}\delta_{ij} - \frac{1}{V_{\mathrm{s}}}\frac{\partial \varphi}{\partial \varepsilon_{ij}} \tag{12.6}$$

where φ is the potential energy, ε_{ij} is a component of the strain tensor, and δ_{ij} is the Kronecker delta function. Depending on the ensemble modelled, any of these quantities will vary in time, fluctuating about a mean value, so for accurate results the simulations need to be performed over a period of time that is large compared with the time scale of the fluctuations. Thermodynamic quantities can be calculated by performing simulations under a range of conditions, for example at different temperatures under conditions of either constant volume or constant pressure. Thus the heat capacity, for example, can be estimated from the difference in energy of the simulation sample between two simulations performed at slightly different temperatures. This approach is similar to the experimental measurement of macroscopic thermodynamic quantities.

The theory of statistical thermodynamics gives relationships between macroscopic thermodynamic quantities and fluctuations of related quantities. For example, the fluctuations in the temperature of a microcanonical ensemble are related to the heat capacity C_V:

$$C_V = \frac{3}{2}Nk_{\mathrm{B}}\left[1 - \frac{3N\left\langle\left(T-\overline{T}\right)^2\right\rangle}{\overline{T}}\right]^{-1} \tag{12.7}$$

where $\overline{T}$ is the mean temperature. This result was derived by Lebowitz et al. (1967). Other important quantities that can be calculated from fluctuations are the compressibility from the pressure fluctuations (Cheung 1977), and the elastic constants from the strain fluctuations in a constant-pressure simulation. We stress again that these quantities will all be calculated classically; a method for calculating quantum thermodynamic quantities from classical simulations is described later.

[5] This may sound odd, but in some cases long-range interactions can be modelled as a volume-dependent constant (Dove 1988).

Structural information

The simulation can give direct information about the crystal structure. For example, the mean position within the unit cell of a given atom can readily be evaluated by averaging over all unit cells over a reasonable period of time. For comparison with X-ray crystallographic data, the mean-square atomic displacement, $\langle u^2 \rangle = \langle | \mathbf{r} - \langle \mathbf{r} \rangle |^2 \rangle$, is a straightforward and useful quantity to calculate. This is one quantity that can be dependent on sample size. The main contribution to the atomic displacements is from the long-wavelength acoustic modes, but in small samples the coarse sampling of reciprocal space means that these lattice vibrations are not present. For example, in a simulation of $MgSiO_3$ perovskite using a large sample, Winkler and Dove (1992) obtained larger values of the anisotropic mean-square displacements than Matsui (1988), although both studies used the same interatomic potential model and the same thermodynamic conditions.

It may often be useful to calculate the intensity for X-ray or neutron scattering, either from Bragg peaks following equation (E.11), or diffuse scattering following equation (E.12). Bounds et al. (1980) used such calculations for solid methane to test a theory of elastic and diffuse scattering from orientationally disordered crystals (Dolling et al. 1979). One advantage of using simulations for the calculation of diffuse scattering is that it is possible to identify different contributions to the overall scattering, allowing a detailed interpretation of experimental data (Dove and Pawley 1984).

Dynamic information: single particle correlation functions

Dynamic information from MDS studies is conveniently obtained by calculations of time correlation functions, as described in Appendix F. If we have a dynamic variable x associated with each atom, the correlation function $C(t)$ for the time $t = n\Delta t$ can be constructed from the general equation:

$$C(t) = \frac{1}{N(M-n)} \sum_{j=1}^{N} \sum_{m=1}^{M-n} x_j(m\Delta t) x_j(m\Delta t + t) \qquad (12.8)$$

where M is the total number of time steps.

The quantity x in equation (12.8) may, for example, be an atomic velocity. The velocity correlation function averaged over all atoms will yield the Fourier transform of the phonon density of states, as described in Appendix F. The phonon density of states is useful for one method of taking quantum-mechanical effects into account in the evaluation of thermodynamic quantities. The fundamental thermodynamic quantities can be calculated from the density of

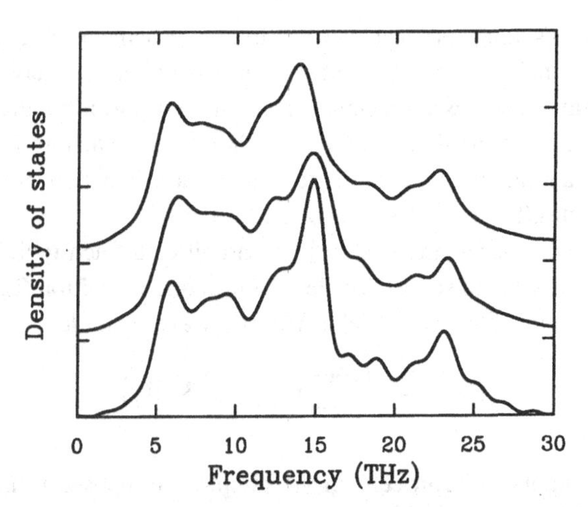

Figure 12.1: The density of states for $MgSiO_3$ perovskite calculated from simulations performed at two different temperatures (Winkler and Dove 1992). The lowest curve is for $T = 300$ K and nominally 1 bar pressure; the middle curve is for $T = 1000$ K and the same cell volume as the lowest curve; the top curve is for $T = 1000$ K and nominally 1 bar pressure (i.e. larger cell volume). These calculations show the shifts in frequency due to thermal expansion (lowering of the frequencies of the prominent peaks in the top curve) and intrinsic anharmonic interactions (raising of the frequencies of the prominent peaks in the middle curve).

states using equations (5.4)–(5.7). This approach can be called a *semi-classical approximation*, in that the density of states $g(\omega)$ is calculated in the simulation classically, but the thermodynamic functions are then calculated quantum-mechanically. This method has recently been applied to an MDS study of $MgSiO_3$ perovskite (Winkler and Dove 1992). It is expected that when anharmonic effects are not very large, the quantum corrections to $g(\omega)$ will be negligibly small. The quantum corrections arise through the difference between the quantum-mechanical or classical calculations of the shifts in phonon frequencies due to anharmonic interactions, equation (8.17). The density of states calculated for $MgSiO_3$ is shown in Figure 12.1.

Dynamic information: phonons and collective fluctuations

For the study of individual phonons we need to consider the Fourier components of the atomic trajectories. One straightforward method is to simulate a scattering experiment by calculating the inelastic scattering function $S(\mathbf{Q}, \omega)$, as given by equation (9.2) (Klein 1978). For simple systems, where there is a clear distinction between transverse and longitudinal modes, and between acoustic and optic modes, this method gives reasonable results. However, for complex systems this approach suffers from the same problems that are encountered in inelastic neutron scattering measurements, namely that it may not yield unambiguous assignments of phonon modes to individual peaks in the spectra, and that the determination of phonon eigenvectors is extremely difficult. The alternative approach is to perform a calculation of the normal mode coordinates (Dove and Lynden-Bell 1986).

Let us first consider a single atom in each unit cell. Each individual atom (labelled j) has a value of some variable x_j, and an origin position $\mathbf{R}_j$, so that we can define a general collective variable $X(\mathbf{k}, t)$ at wave vector $\mathbf{k}$:

$$X(\mathbf{k},t) = \frac{1}{\sqrt{N}} \sum_j x_j(t) \exp\left(i\mathbf{k} \cdot \mathbf{R}_j\right) \tag{12.9}$$

N is now the number of unit cells in the sample (as opposed to its previous definition as the total number of atoms in the sample). For the phonon normal mode coordinates, the variable x_j will be equal to the displacement of the atom from its average position along one of the principal axes, u_j, multiplied by the square root of the atomic mass, m_j:

$$U(\mathbf{k},t) = \frac{1}{\sqrt{N}} \sum_j \sqrt{m_j}\, u_j(t) \exp\left(i\mathbf{k} \cdot \mathbf{R}_j\right) \tag{12.10}$$

We need to extend this by defining collective variables for each of the displacements along the three principal axes for each atom in the unit cell, giving a total of $3Z$ collective variables (where Z is the number of atoms in each unit cell). In the case of a molecular crystal, we would define three displacement variables and three orientational variables for each molecule. For the orientational case the mass should be replaced by the relevant component of the inertia tensor, given that the molecular axes are chosen in such a way that the inertia tensor is diagonal.

Two comments on the practical application of equation (12.10) can be made. First, the displacements can be replaced by the actual positions (relative to the unit cell origin) for the computation of $U(\mathbf{k}, t)$. We can show that this works by defining the displacement $u(t)$ as being equal to the difference between the instantaneous position, $r(t)$, and the mean position, $\bar{r}$.[6] We can then write the relevant part of equation (12.10) as

$$\sum_j u_j(t) \exp\left(i\mathbf{k} \cdot \mathbf{R}_j\right) = \sum_j r_j(t) \exp\left(i\mathbf{k} \cdot \mathbf{R}_j\right) - \sum_j \bar{r}_j \exp\left(i\mathbf{k} \cdot \mathbf{R}_j\right) \tag{12.11}$$

[6] For the moment we preserve the distinction between $\bar{\mathbf{r}}$ and $\mathbf{R}$, although technically these may be equivalent.

For wave vectors that are not also reciprocal lattice vectors, the second term on the left hand side of equation (12.11) is equal to zero. The second comment is that the value of $\mathbf{R}_j$ in equation (12.10) can be replaced by the position of the origin of the unit cell containing the particular atom. This difference merely adds a phase factor to the equation, which is lost in subsequent analysis. The advantage of using the unit cell origin is that there is then no need to determine accurately the average position of the atom.

We now need to see how the $3Z$ collective variables given by equation (12.10) can be converted to normal mode coordinates. First we gather the $3Z$ collective variables for each time step into a single $3Z$-component column vector $\mathbf{T}(\mathbf{k}, t)$. The set of normal mode coordinates is similarly defined in terms of a $3Z$-component column vector $\mathbf{Q}(\mathbf{k}, t)$. These two vectors are related by a simple transformation equation:

$$\mathbf{Q}(\mathbf{k}, t) = \mathbf{A}(\mathbf{k}) \cdot \mathbf{T}(\mathbf{k}, t) \tag{12.12}$$

where $\mathbf{A}(\mathbf{k})$ is a $3Z \times 3Z$ matrix with time-independent components. $\mathbf{A}(\mathbf{k})$ is determined by calculating the following matrix equation for the thermal and time averages:

$$\left\langle \mathbf{Q}(\mathbf{k}) \cdot \mathbf{Q}^{\mathrm{T}}(-\mathbf{k}) \right\rangle = \mathbf{A}(\mathbf{k}) \cdot \left\langle \mathbf{T}(\mathbf{k}) \cdot \mathbf{T}^{\mathrm{T}}(-\mathbf{k}) \right\rangle \cdot \mathbf{A}^{\mathrm{T}}(\mathbf{k}) = \mathbf{A} \cdot \mathbf{S} \cdot \mathbf{A}^{\mathrm{T}} \tag{12.13}$$

where the superscript T denotes the transpose, and equation (12.13) defines the matrix $\mathbf{S}(\mathbf{k})$. $\mathbf{S}(\mathbf{k})$ is the quantity that is calculated over a period of time in a simulation. Since the normal mode coordinates are orthogonal, the left hand side of equation (12.13) is a diagonal matrix. Thus equation (12.13) represents a matrix diagonalisation procedure, and the matrix $\mathbf{A}(\mathbf{k})$ is defined by the eigenvectors of $\mathbf{S}(\mathbf{k})$. After using the simulation data to generate the matrix $\mathbf{A}(\mathbf{k})$, the time-dependent normal mode coordinates can easily be constructed from equation (12.11). The matrix $\mathbf{A}(\mathbf{k})$ contains the set of normal mode eigenvectors.

Having used this procedure to generate the time-dependent normal mode coordinates, the frequencies of the normal modes can be obtained by two methods. Firstly, if $Q(\mathbf{k}, \nu, t)$ is the normal mode coordinate for the phonon mode labelled ν, the corresponding phonon frequency can be calculated using the standard classical quasi-harmonic result (equation (4.19)):

$$\omega^2(\mathbf{k}, \nu) = \frac{k_{\mathrm{B}}T}{\left\langle Q(\mathbf{k}, \nu) Q(-\mathbf{k}, \nu) \right\rangle} \tag{12.14}$$

where the denominator is given by the eigenvalues of $\mathbf{S}(\mathbf{k})$, equation (12.13). This method is in principle fairly straightforward, and involves less effort than the second method described below. It is, however, not as accurate as the second method, because it assumes that there is complete equipartition of energy between all the normal modes. This may not be so, particularly if anharmonic effects are weak and the transfer of energy between different normal modes is slow compared with the period of time sampled by the simulation. This limitation is most severe for the acoustic modes.[7] In part this problem can be circumvented by replacing equation (12.14) with:

$$\omega^2(\mathbf{k},\nu) = \frac{\langle \dot{Q}(\mathbf{k},\nu)\dot{Q}(-\mathbf{k},\nu)\rangle}{\langle Q(\mathbf{k},\nu)Q(-\mathbf{k},\nu)\rangle} \tag{12.15}$$

The momentum variables can be calculated using a formula analogous to equation (12.12), where the matrix corresponding to $\mathbf{T}$ is constructed from the velocities rather than the positions, and the matrix $\mathbf{A}$ is the same as determined from the positions.

The second method is to use the time dependence of the normal mode coordinates to construct the dynamic correlation function $\langle Q(\mathbf{k},\nu,t)Q(-\mathbf{k},\nu,0)\rangle$. The frequency can be read from the period of the correlation function, or the correlation function can be Fourier transformed (directly, or using the Wiener–Khintchine method described in Appendix F). There should be a single peak in the Fourier transform, with a mid-point frequency corresponding to the phonon frequency. An alternative approach to Fourier transformation, which overcomes problems associated with truncation of the correlation function at a time less than the lifetime of the correlation function, is to fit the correlation function with a damped oscillator function. This will yield both the frequency and the mode lifetime. The frequency obtained from the time correlation function is not affected by the assumption of equipartition. However, for complex structures the effort involved in analysing a large number of correlation functions may be less tolerable than the limitations on the accuracy imposed using equation (12.14)!

The practical implementation of equation (12.13) can be simplified by incorporating the constraints of symmetry. Without these constraints, the symmetry of the normal modes will not be completely retained owing to errors associated with sampling only a short time during the simulation. If we include symmetry, we automatically ensure that the normal modes have the correct symmetry as

[7] Dove et al. (1986) give an analysis of an acoustic mode with a correlation function that has a significant imaginary component which theoretically should be zero.

well as reducing the computational effort. Let us consider a crystal with four identical atoms in the unit cell, and consider only the motions along the x-axis. We construct the following four collective variables:

$$X_1(\mathbf{k}) = \sqrt{\frac{m}{N}} \sum_j \left[x_1(j) + x_2(j) + x_3(j) + x_4(j)\right] \exp\left(i\mathbf{k} \cdot \mathbf{R}_j\right) \quad (12.16a)$$

$$X_2(\mathbf{k}) = \sqrt{\frac{m}{N}} \sum_j \left[x_1(j) - x_2(j) + x_3(j) - x_4(j)\right] \exp\left(i\mathbf{k} \cdot \mathbf{R}_j\right) \quad (12.16b)$$

$$X_3(\mathbf{k}) = \sqrt{\frac{m}{N}} \sum_j \left[x_1(j) - x_2(j) - x_3(j) + x_4(j)\right] \exp\left(i\mathbf{k} \cdot \mathbf{R}_j\right) \quad (12.16c)$$

$$X_4(\mathbf{k}) = \sqrt{\frac{m}{N}} \sum_j \left[x_1(j) + x_2(j) - x_3(j) - x_4(j)\right] \exp\left(i\mathbf{k} \cdot \mathbf{R}_j\right) \quad (12.16d)$$

These collective variables are used instead of the raw coordinates to form the matrix **S**(**k**) in equation (12.13). Depending on the symmetry of the crystal structure, these combinations will preserve some symmetry elements of the structure but will break others; each of the four combinations will therefore be defined as a particular symmetry. The analogy between equations (12.16a–d) and Table 3.1 should be noted. Detailed application will require a knowledge of the symmetry of the normal modes for a given wave vector. Although the incorporation of symmetry is not essential, it is desirable since we can construct small versions of equation (12.13) for each symmetry, rather than one overall equation for all modes. Moreover, the symmetry gives an unambiguous identification of a calculated frequency with a given branch in the dispersion curves, particularly since the accuracy of the calculated frequencies may not be high.

The dispersion curves have been calculated for β-quartz for **k** along [100] by MDS by Tautz et al. (1991) using the model of Tsuneyuki et al. (1988, 1990). In this example, there are 27 normal modes but the point symmetry of the wave vector contains only two irreducible representations. The use of symmetry constraints did not significantly affect the calculated frequencies, but the unambiguous assignment of all the modes was impossible without the symmetry constraints. This study of β-quartz highlights the value of the MDS method. Because the β phase is unstable at low temperatures, harmonic lattice dynamics calculations for the β phase will give some imaginary branches, reflecting the fact that the structure does not correspond to a minimum of the potential energy. Thus analysis of the collective excitations of the β phase cannot be obtained by any method other than MDS. A detailed analysis of the soft mode in β-quartz calculated in the simulation sheds light on the fundamental interactions that give rise to the incommensurate phase transition.

Model systems

Our discussion of the MDS method has focussed on the use of realistic model interatomic potentials. It is sometimes useful to strip a physical system to its bare bones, by inventing a model potential that contains the minimum number of features essential to give the desired behaviour. In this way it is possible to study the phenomena of interest without having to worry about other complications, and it is then easier to relate the observed behaviour with specific features of the interactions. For example, the system described by equation (8.29), where the phenomenon of interest is the second-order phase transition, has been studied in some detail by MDS. An extensive set of simulations under various conditions has revealed many complex properties close to the transition, such as cluster formation and the observation of the development of a central peak in the power spectra at temperatures close to the transition temperature (Schneider and Stoll 1973, 1975, 1976, 1978). Giddy et al. (1990) have used similar simulations to give comparisons with theoretical calculations of Landau free energy functions, and to investigate order–disorder phenomena (Normand et al. 1990). Parlinski (1988) has used different models to give information about incommensurate ordering. Such calculations have prompted the use of the term *computer experiment* to describe the approach of using simulations to perform experiments with the fundamental equations of motion.

Limitations of the molecular dynamics simulation method

So why bother setting up complex experiments if the same information can be obtained using molecular dynamics simulations? This question is answered by noting that the MDS method contains inherent limitations, which we can list:

1 The computational demands limit the range of conditions over which simulations can be performed. An experiment may be performed at a number of different temperatures, whereas it is often only practical to perform detailed simulations at only a few temperatures.
2 Simulations are limited by the accuracy of the interatomic potential being used, the sophistication of which is limited by the computational resources as much as by the need to use a mathematical representation of the actual

potential. For example, it is generally appreciated that ionic polarisability is often an important component of the energy of the crystal, yet it is very difficult to incorporate shell models into MDS calculations. Hence most simulations are restricted to the use of the rigid-ion model.

3 The use of finite-sized samples and periodic boundary conditions also imposes severe limitations. One example is that a travelling wave will traverse the sample in a short time, and the wavefront will reappear at the same point. The interaction of this reappearing wave with the origin of the wave is not understood. The use of finite sizes restricts the range of allowed wave vectors. This is unfortunate if wave vector dependent behaviour is to be studied, and requires the use of large samples if detailed analysis is required. However, large samples generally require longer running times owing to the existence of the long-wavelength acoustic modes, and hence require even greater computational resources. The use of a restricted set of wave vectors has an even more fundamental problem, in that small samples may give different results from large samples. For example, the mean-squared atomic displacement given by equation (4.21) involves a sum over all normal modes. For a restricted set of wave vectors, many of the wave vectors that exist in a macroscopic crystal will be absent. The important modes for the displacement amplitude are the long-wavelength acoustic modes, since the contribution of any mode is weighted by the inverse of the square of the mode frequency. Most of these acoustic modes are absent in a simulation using a small sample, and so in general the atomic displacement amplitude calculated in a simulation will be an underestimate of the true amplitude. This point has been demonstrated by Winkler and Dove (1992), as discussed above.

4 The fact that a simulation can only be run for a length of time that is tiny compared to the time scale of a normal experiment also presents a problem, namely that the sample does not have enough time to evolve through the whole of the multidimensional phase space in a true ergodic manner. This means that the results will surely be biased by the starting conditions and the length of time the simulation is run for. This bias will not generally give wrong results, but will be reflected in the accuracy of the results. The calculations of phonon dispersion curves highlight this factor.

Despite these limitations, however, it is often the case that a good MDS calculation performed in conjunction with experimental and theoretical work is able to provide insights that cannot be obtained by any other method, since the limitations of the technique are quite different from the limitations experienced with experiment or theory.

Summary

1. The molecular dynamics simulation method gives a direct simulation of the evolution with time of a small ensemble of atoms that obey classical mechanics.
2. The continuous equations of motion are modelled using time-step algorithms.
3. A number of different thermodynamic ensembles can be simulated.
4. Simulations are able to give macroscopic thermodynamic data, structural data, and dynamic data.
5. The methods for extracting normal mode coordinates, eigenvectors, and frequencies from a simulation are described.

FURTHER READING

Allen and Tildesley (1987)
Ciccotti et al. (1987)

Appendix A
The Ewald method

The Ewald method is a technique for the summation of long-range interactions. The usual application, namely for the evaluation of the Coulomb energy of a crystal, is described. The modifications that enable the technique to be used for other inverse law forces are outlined using the specific example of the dispersive r^{-6} interaction.

The Ewald sum for the Coulomb energy

The contribution of the Coulomb energy to the lattice energy is extremely difficult to evaluate by merely summing over enough neighbouring ion pairs because the summation converges too slowly for accurate results to be practicably obtained. Ewald (1921) developed a solution to this problem by noting that the summation can be split into two separate summations, one in real space and one in reciprocal space, which are both rapidly convergent. The starting point is to note that $1/r$ can be written as

$$\frac{1}{r} = \frac{2}{\sqrt{\pi}} \int_0^{\infty} \exp\left(-r^2 \rho^2\right) \mathrm{d}\rho \tag{A.1}$$

The Coulomb contribution to the lattice energy of the crystal, W_C, can then be written as

$$W_C = \frac{1}{2} \sum_{l=0}^{\infty} \sum_{i,j} \frac{Q_i Q_j}{4\pi\varepsilon_0 r_{ij}(l)} = \frac{1}{2} \sum_{l=0}^{\infty} \sum_{i,j} \frac{Q_i Q_j}{2\pi^{3/2}\varepsilon_0} \int_0^{\infty} \exp\left(-r_{ij}^2(l)\rho^2\right) \mathrm{d}\rho \tag{A.2}$$

where $r_{ij}(l)$ is the distance between the i-th atom in the reference unit cell (labelled 0) and the j-th atom in the l-th unit cell. Q_j is the charge on the atom. The integral can then be split into two halves:

$$\int_0^\infty \exp(-r^2\rho^2)\mathrm{d}\rho = \int_0^g \exp(-r^2\rho^2)\mathrm{d}\rho + \int_g^\infty \exp(-r^2\rho^2)\mathrm{d}\rho$$
$$= \int_0^g \exp(-r^2\rho^2)\mathrm{d}\rho + \frac{\sqrt{\pi}}{2}\frac{\mathrm{erfc}(gr)}{r} \qquad \text{(A.3)}$$

where g is a free parameter, and erfc is the *complementary error function*:

$$\mathrm{erfc}(x) = 1 - \mathrm{erf}(x) = \frac{2}{\sqrt{\pi}}\int_x^\infty \exp(-y^2)\mathrm{d}y \qquad \text{(A.4)}$$

The complementary error function falls to zero quite quickly on increasing r, and so the second component of the right hand side of equation (A.3) needs no further treatment before it is linked in with equation (A.2). Although the first term on the right hand side of equation (A.3) is still a slow function of r, we can note that an expression that falls to zero slowly in real space will fall to zero quickly if it is transformed into reciprocal space. In Appendix B we obtain the relation:

$$\frac{2}{\sqrt{\pi}}\sum_l \exp(-r_{ij}^2(l)\rho^2) = \frac{2\pi}{V_\mathrm{c}}\sum_{\mathbf{G}} \rho^{-3}\exp(-G^2/4\rho^2)\exp(i\mathbf{G}\cdot(\mathbf{r}_j - \mathbf{r}_i)) \qquad \text{(A.5)}$$

where V_c is the volume of the unit cell, and $\mathbf{G}$ is a reciprocal lattice vector. In equation (A.5) the atomic positions $\mathbf{r}_i$ and $\mathbf{r}_j$ are both within the reference unit cell. Equation (A.5) is known as Ewald's *theta function transformation*. Bringing together equations (A.2), (A.3) and (A.5) gives:

$$W_\mathrm{C} = \frac{1}{2}\sum_{l=0}^{\infty}\sum_{i,j}\frac{Q_iQ_j}{4\pi\varepsilon_0}\frac{\mathrm{erfc}(gr_{ij}(l))}{r_{ij}(l)}$$
$$+\frac{1}{2}\sum_{i,j}\frac{Q_iQ_j}{4\pi\varepsilon_0}\frac{2\pi}{V_\mathrm{c}}\int_0^g\sum_{\mathbf{G}}\rho^{-3}\exp(-G^2/4\rho^2)\exp(i\mathbf{G}\cdot(\mathbf{r}_j-\mathbf{r}_i))\mathrm{d}\rho \qquad \text{(A.6)}$$

We now note the result of the integral:

$$\int_0^g \rho^{-3}\exp(-G^2/4\rho^2)\exp(i\mathbf{G}\cdot(\mathbf{r}_j-\mathbf{r}_i))\mathrm{d}\rho$$
$$= \frac{2\exp(-G^2/4g^2)}{G^2}\exp(i\mathbf{G}\cdot(\mathbf{r}_j-\mathbf{r}_i)) \qquad \text{(A.7)}$$

Therefore the final expression for the Coulomb energy is given as

$$W_{\mathrm{C}} = \frac{1}{2}\sum_{l=0}^{\infty}\sum_{i,j}\frac{Q_i Q_j}{4\pi\varepsilon_0}\frac{\mathrm{erfc}\left(g r_{ij}(l)\right)}{r_{ij}(l)}$$

$$+\frac{1}{2}\sum_{i,j}\frac{Q_i Q_j}{4\pi\varepsilon_0}\frac{4\pi}{V_{\mathrm{c}}}\sum_{\mathbf{G}}\frac{\exp\left(-G^2/4g^2\right)}{G^2}\exp\left(i\mathbf{G}\cdot\left(\mathbf{r}_j-\mathbf{r}_i\right)\right) \qquad \text{(A.8)}$$

We need to note two points concerning equation (A.8). The first is that we have implicitly included the interactions for $i = j$ when $l = 0$; this is required for the reciprocal space transformation to work. We therefore need to subtract these terms from equation (A.8), which occur in both the real and reciprocal space summations. This is accomplished with an additional term, known as the *self term* W_{self}:

$$W_{\mathrm{self}} = \lim(r \to 0)\frac{1}{2}\frac{1}{4\pi\varepsilon_0}\sum_i \frac{Q_i^2\left[\mathrm{erfc}(gr)-1\right]}{r} \qquad \text{(A.9)}$$

We use the small argument expansion:

$$1-\mathrm{erfc}(x) = \mathrm{erf}(x) \;\; ; \;\; \lim(x \to 0)\,\mathrm{erf}(x) = \frac{2x}{\sqrt{\pi}} \qquad \text{(A.10)}$$

The self term is therefore given by

$$W_{\mathrm{self}} = -\frac{1}{4\pi\varepsilon_0}\sum_i \frac{gQ_i^2}{\sqrt{\pi}} \qquad \text{(A.11)}$$

The second point about the reciprocal space term is that it includes the contribution for $\mathbf{G} = 0$, which needs to be taken out of the summation. In fact this term is discarded from the final energy expression, for the following reasons. The contribution for $\mathbf{G} = 0$ can be written as

$$\lim(\mathbf{G} \to 0)\frac{\exp\left(-G^2/4g^2\right)}{G^2}\left|F_Q(\mathbf{G})\right|^2 \qquad \text{(A.12)}$$

where we have dropped the constant factors, and the structure factor $F_Q(\mathbf{G})$ has the form

$$F_Q(\mathbf{G}) = \sum_j Q_j \exp\left(i\mathbf{G}\cdot\mathbf{r}_j\right) \qquad \text{(A.13)}$$

In the limit $\mathbf{G} \to 0$ we can write $F_Q(\mathbf{G})$ as

$$\lim(\mathbf{G} \to 0) F_Q(\mathbf{G}) \approx \sum_j Q_j \left(1 + i\mathbf{G} \cdot \mathbf{r}_j\right)$$

$$= i\mathbf{G} \cdot \sum_j Q_j \mathbf{r}_j = i\mathbf{G} \cdot \boldsymbol{\mu} \qquad \text{(A.14)}$$

where $\boldsymbol{\mu}$ is the dipole moment of the unit cell, and we have used the fact that the sum of the charges within a unit cell is zero. If the unit cell is centrosymmetric, $\boldsymbol{\mu} = 0$,[1] and hence $F_Q(0)$ will be zero. In fact $|F_Q(0)|^2$ becomes the dominant component in the $\mathbf{G} = 0$ term, which therefore has a value of zero. For the case where $\boldsymbol{\mu}$ is non-zero, we write $\mathbf{G}$ as $G\mathbf{e}$, where $\mathbf{e}$ is a unit vector in the direction of $\mathbf{G}$, so that the $\mathbf{G} = 0$ term becomes simply $[\mathbf{e} \cdot \boldsymbol{\mu}]^2$. Thus the value of the $\mathbf{G} = 0$ term is dependent on the direction from which $\mathbf{G}$ approaches zero. Because of this ambiguity we do not include this term in the Coulomb energy, equation (A.8). Instead we interpret the $\mathbf{G} = 0$ term as a macroscopic applied field.

The $\mathbf{G} = 0$ term is of interest as it highlights the concept of LO/TO splitting introduced in Chapter 3. The energy of a mode of wave vector $\mathbf{k}$ in the limit $\mathbf{k} \to 0$ that generates a dipole moment $\boldsymbol{\mu}$ in the unit cell (such as both the LO and TO modes in NaCl) will depend on the value of $\mathbf{k} \cdot \boldsymbol{\mu}$. This will have different values for LO modes, in which $\boldsymbol{\mu}$ will be parallel to $\mathbf{k}$, and TO modes, in which $\boldsymbol{\mu}$ will be perpendicular to $\mathbf{k}$.

We therefore conclude that the Coulomb sum can be written in the final form:

$$W_C = \frac{1}{2} \sum_{l=0}^{\infty} \sum_{i,j} \frac{Q_i Q_j}{4\pi\varepsilon_0} \frac{\operatorname{erfc}\left(g r_{ij}(l)\right)}{r_{ij}(l)}$$

$$+ \frac{1}{2} \sum_{i,j} \frac{Q_i Q_j}{4\pi\varepsilon_0} \frac{4\pi}{V_c} \sum_{\mathbf{G} \neq 0} \frac{\exp\left(-G^2/4g^2\right)}{G^2} \exp\left(i\mathbf{G} \cdot \left(\mathbf{r}_j - \mathbf{r}_i\right)\right) + W_{\text{self}} \qquad \text{(A.15)}$$

where, it should be recalled, the first summation does not include the terms for $i = j$ when $l = 0$.

The parameter g has been included as a parameter that can be chosen to have any arbitrary value. In practice the exact value controls the convergence of both the real and reciprocal space summations; a large value of g gives rapid convergence of the real space summation, and a small value gives rapid convergence of the reciprocal space summation. Given that the number of computations in each term varies as the third power of the cut-off limit, a good compromise value of g will ensure that both summations converge with more-or-

[1] $\boldsymbol{\mu}$ may also have a value of zero in non-centrosymmetric crystals.

less equal numbers of terms. Fine tuning of the exact value of g will allow computer programs to run as fast as possible.

It also needs to be noted that for molecular crystals, we are usually only interested in the *intermolecular* interactions. The Ewald sum implicitly includes all the *intramolecular* interactions, just as it included the interactions of ions with themselves (the self terms). The intramolecular interactions will therefore need to be subtracted from equation (A.15) to give the correct lattice energy.

Extension for other terms of the functional form r^{-n}: the case $n = 6$

Any interaction that varies as r^{-n} can be summed using a modification of the Ewald method.[2] We specifically consider the case $n = 6$, appropriate for the dispersive interaction. Although this interaction falls much faster with r than the Coulomb interaction, truncation of the summation at some practical cut-off limit will always introduce a residual error. We note the general result:

$$\frac{1}{r^{2n}} = \int_0^\infty \rho^{2n-1} \exp\left(-r^2\rho^2\right) \mathrm{d}\rho$$
$$\Rightarrow \quad \frac{1}{r^6} = \int_0^\infty \rho^5 \exp\left(-r^2\rho^2\right) \mathrm{d}\rho \qquad \text{(A.16)}$$

The dispersive contribution to the lattice energy, W_D, can be written as

$$W_\mathrm{D} = -\frac{1}{2} \sum_{l=0}^{\infty} \sum_{i,j} A_{ij} r_{ij}^{-6}(l) \qquad \text{(A.17)}$$

We follow the procedure outlined above and note that the integral in equation (A.16) can be split into two halves when substituted into equation (A.17).

The integral for the real space term has the solution:

$$\int_g^\infty \rho^5 \exp\left(-r^2\rho^2\right) \mathrm{d}\rho = \frac{\exp\left(-r^2 g^2\right)}{r^2} \left[\frac{g^4}{2} + \frac{g^2}{r^2} + \frac{1}{r^4} \right] \qquad \text{(A.18)}$$

We use the theta function transformation given in equation (A.5) to obtain the reciprocal lattice sum:

[2] The method described here was pointed out to the author by Alastair I. M. Rae.

$$-\frac{1}{2}\sum_{l=0}^{\infty}\sum_{i,j} A_{ij}\int_0^g \rho^5 \exp\left(-r_{ij}^2(l)\rho^2\right)\mathrm{d}\rho$$

$$= -\frac{\pi^{3/2}}{2V_\mathrm{c}}\sum_{\mathbf{G}}\sum_{i,j} A_{ij}\int_0^g \rho^2 \exp\left(-G^2/4\rho^2\right)\exp\left(i\mathbf{G}\cdot\left(\mathbf{r}_j-\mathbf{r}_i\right)\right)\mathrm{d}\rho \quad \text{(A.19)}$$

The solution to the integral in equation (A.19) is given as

$$\int_0^g \rho^2 \exp\left(-G^2/4\rho^2\right)\mathrm{d}\rho$$

$$= \frac{G^2 g}{6}\exp\left(-G^2/4g^2\right)\left[\frac{2g^2}{G^2}-1\right]+\frac{\sqrt{\pi}G^3}{12}\operatorname{erfc}(G/2g) \quad \text{(A.20)}$$

The term for $\mathbf{G}=0$ is given as

$$W_\mathrm{D}(\mathbf{G}=0) = -\frac{g^3\pi^{3/2}}{6V_\mathrm{c}}\sum_{i,j} A_{ij} \quad \text{(A.21)}$$

The self term (i.e. the term in the real space sum with $i=j$ when $l=0$) is given as

$$W_\mathrm{self} = \frac{g^2}{12}\sum_i A_{ii} \quad \text{(A.22)}$$

When we add together equations (A.17)–(A.22), we obtain the final expression for the dispersive contribution to the lattice energy:

$$W_\mathrm{D} = -\frac{1}{2}\sum_{l=0}^{\infty}\sum_{i,j} A_{ij} r_{ij}^{-6}(l)\exp\left(-g^2 r_{ij}^2(l)\right)\left[1+g^2 r_{ij}^2(l)+\frac{g^4 r_{ij}^4(l)}{2}\right]$$

$$-\frac{\pi^{3/2}}{12V_\mathrm{c}}\sum_{\mathbf{G}\neq 0}\left\{\left[g\exp\left(-G^2/4g^2\right)\left(2g^2-G^2\right)+\frac{\sqrt{\pi}}{2}G^3\operatorname{erfc}(G/2g)\right.\right.$$

$$\left.\left.\times\sum_{i,j} A_{ij}\exp\left(i\mathbf{G}\cdot\left(\mathbf{r}_j-\mathbf{r}_i\right)\right)\right\}-\frac{g^3\pi^{3/2}}{6V_\mathrm{c}}\sum_{i,j} A_{ij}+W_\mathrm{self} \quad \text{(A.23)}$$

Equation (A.23) was first derived by Williams (1971), using a somewhat different approach from that outlined here. In general use the sum has to be performed for every specific interaction type. Williams points out that if the coefficients in the dispersive interaction are subject to the reasonable constraint:

$$A_{ij} = \sqrt{A_{ii}A_{jj}} \tag{A.24}$$

the summation is easier to implement.

We note that the choice of the value of g will probably not be the same for the dispersive lattice energy as for the Coulomb lattice energy. However, the same criterion holds, namely that the best value of g will ensure rapid convergence of both the real and reciprocal space sums, reducing the computing time to a minimum.

Finally, the comments given above concerning the inclusion of intramolecular interactions apply to the dispersive energy in the same way as for the Coulomb energy.

FURTHER READING

Born and Huang (1954) ch. 30, App. II–III
Brüesch (1982) App. K

Appendix B
Lattice sums

The various lattice sums that are used in this book are evaluated.

Two fundamental results

All the results derived in this appendix will rely on two fundamental results that apply to crystals containing N unit cells with periodic boundary conditions:

$$\frac{1}{N}\sum_{\mathbf{R}}\exp(i\mathbf{k}\cdot\mathbf{R})=\delta_{\mathbf{k},0} \tag{B.1}$$

$$\frac{1}{N}\sum_{\mathbf{k}}\exp(i\mathbf{k}\cdot\mathbf{R})=\delta_{\mathbf{R},0} \tag{B.2}$$

where $\mathbf{k}$ is a wave vector in the first Brillouin zone, and $\mathbf{R}$ is a lattice vector equal to $u\mathbf{a}+v\mathbf{b}+w\mathbf{c}$ (u, v, and w are integers).

Equations (B.1) and (B.2) can readily be derived mathematically (e.g. Squires 1978, App. A4; Brüesch 1982, App. B). Ashcroft and Mermin (1976, App. F) point out that these equations can be deduced to be true by simple reasoning. When periodic boundaries are assumed, the sum of equation (B.1) should be independent of the origin of the sample. This means that the same result should be obtained when an arbitrary lattice vector is added to every value of $\mathbf{R}$. For this to be true, $\mathbf{k}$ has to be equal to 0 (or to a reciprocal lattice vector $\mathbf{G}$) unless the sum is to have the trivial result of 0; clearly when $\mathbf{k}=0$ the result is unity. Similarly the sum in equation (B.2) should be invariant when an arbitrary wave vector is added to every wave vector in equation (B.2). This can only be true when $\mathbf{R}=0$.

Derivation of equation (4.13)

The velocity of an atom is given as

$$\dot{\mathbf{u}}(jl) = \frac{1}{(Nm_j)^{1/2}} \sum_{\mathbf{k},\nu} \mathbf{e}(j,\mathbf{k},\nu) \exp(i\mathbf{k}\cdot\mathbf{r}(l)) \dot{Q}(\mathbf{k},\nu) \tag{B.3}$$

where for simplicity we take the origin of each atom in the unit cell to be at the same point as all others. The kinetic energy of a single atom is therefore given as

$$\frac{1}{2} m_j |\dot{\mathbf{u}}(jl)|^2 = \frac{1}{2N} \sum_{\mathbf{k},\nu} \sum_{\mathbf{k}',\nu'} \{ \mathbf{e}(j,\mathbf{k},\nu) \cdot \mathbf{e}^*(j,\mathbf{k}',\nu') \exp(i(\mathbf{k}-\mathbf{k}')\cdot\mathbf{r}(l)) \times \dot{Q}(\mathbf{k},\nu)\dot{Q}^*(\mathbf{k}',\nu') \} \tag{B.4}$$

In order to obtain the kinetic energy of the whole crystal, we need to sum equation (B.4) over all atoms:

$$\frac{1}{2} \sum_{j,l} m_j |\dot{\mathbf{u}}(jl)|^2 = \frac{1}{2N} \sum_{l} \sum_{\mathbf{k},\nu} \sum_{\mathbf{k}',\nu'} \{ \mathbf{e}(\mathbf{k},\nu) \cdot \mathbf{e}^*(\mathbf{k}',\nu') \exp(i(\mathbf{k}-\mathbf{k}')\cdot\mathbf{r}(l)) \times \dot{Q}(\mathbf{k},\nu)\dot{Q}^*(\mathbf{k}',\nu') \} \tag{B.5}$$

where the mode eigenvector **e** has subsumed all the components of each atom in the unit cell to form a $3Z$-component vector. From equation (B.1) we note that the only terms that do not sum to zero are those for $\mathbf{k} = \mathbf{k}'$. We also note that in this case we lose the dependence on **r**, so the sum over all unit cells is replaced by a factor of N:

$$\begin{aligned} \frac{1}{2} \sum_{j,l} m_j |\dot{\mathbf{u}}(jl)|^2 &= \frac{1}{2N} \sum_{\mathbf{k},\nu,\nu'} \mathbf{e}(\mathbf{k},\nu) \cdot \mathbf{e}^*(\mathbf{k},\nu') \dot{Q}(\mathbf{k},\nu)\dot{Q}^*(\mathbf{k},\nu') \\ &= \frac{1}{2} \sum_{\mathbf{k},\nu} |\dot{Q}(\mathbf{k},\nu)|^2 \end{aligned} \tag{B.6}$$

This result is used in equation (6.31). We note the result for harmonic normal modes:

$$\dot{Q}(\mathbf{k},\nu) = i\omega_\nu(\mathbf{k}) Q(\mathbf{k},\nu) \tag{B.7}$$

which on substitution into equation (B.6) gives the final result:

$$\frac{1}{2} \sum_{j,l} m_j |\dot{\mathbf{u}}(jl)|^2 = \frac{1}{2} \sum_{\mathbf{k},\nu} \omega_\nu^2(\mathbf{k}) |\dot{Q}(\mathbf{k},\nu)|^2 \tag{B.8}$$

Derivation of equation (4.20)

The instantaneous displacement of an atom is given from equation (4.2) as

$$\mathbf{u}(jl) = \frac{1}{\left(Nm_j\right)^{1/2}} \sum_{\mathbf{k},\nu} \mathbf{e}(j,\mathbf{k},\nu)\exp\left(i\mathbf{k}\cdot\mathbf{r}(jl)\right)Q(\mathbf{k},\nu) \tag{B.9}$$

The square of the displacement is therefore given as

$$\left|\dot{\mathbf{u}}(jl)\right|^2 = \frac{1}{Nm_j} \sum_{\mathbf{k},\nu} \sum_{\mathbf{k}',\nu'} \left\{\mathbf{e}(\mathbf{k},\nu)\cdot\mathbf{e}^*(\mathbf{k}',\nu')\exp\left(i(\mathbf{k}-\mathbf{k}')\cdot\mathbf{r}(l)\right)\right. \\ \left. \times Q(\mathbf{k},\nu)Q^*(\mathbf{k}',\nu')\right\} \tag{B.10}$$

By using the same reasoning that led from equation (B.4) to equation (B.6) we obtain the result:

$$\sum_{j,l} \left|\mathbf{u}(jl)\right|^2 = \frac{1}{m_j} \sum_{\mathbf{k},\nu} \left|Q(\mathbf{k},\nu)\right|^2 \tag{B.11}$$

When we replace the sum over all unit cells by an average over all unit cells, we obtain the final result:

$$\left\langle \left|\mathbf{u}(j)\right|^2 \right\rangle = \frac{1}{Nm_j} \sum_{\mathbf{k},\nu} \left\langle \left|Q(\mathbf{k},\nu)\right|^2 \right\rangle \tag{B.12}$$

Derivation of equation (6.44)

The kinetic energy term in equation (6.44) was derived above, equation (B.6). The potential energy term, V, using normal mode coordinates is obtained from the potential energy expressed in terms of atomic displacements, equation (6.11):

$$V = \frac{1}{2} \sum_{jj',ll'} \mathbf{u}^{\mathrm{T}}(jl)\cdot\mathbf{\Phi}\begin{pmatrix} jl \\ j'l' \end{pmatrix}\cdot\mathbf{u}(j'l') \tag{B.13}$$

Substitution for the displacements (B.9) into equation (B.13) yields:

$$\mathbf{u}^{\mathrm{T}}(jl)\cdot\mathbf{\Phi}\begin{pmatrix}jj'\\ ll'\end{pmatrix}\cdot\mathbf{u}(j'l')$$

$$=\frac{1}{N\left(m_j m_{j'}\right)^{1/2}}\sum_{\mathbf{k},\nu}\sum_{\mathbf{k}',\nu'}\left\{\mathbf{e}^{\mathrm{T}}(j,\mathbf{k},\nu)\cdot\mathbf{\Phi}\begin{pmatrix}jj'\\ ll'\end{pmatrix}\cdot\mathbf{e}(j',\mathbf{k}',\nu')\right.$$

$$\left.\times\exp\left(i\left[\mathbf{k}\cdot\mathbf{r}(jl)+\mathbf{k}'\cdot\mathbf{r}(j'l')\right]\right)Q(\mathbf{k},\nu)Q(\mathbf{k}',\nu')\right\} \qquad \text{(B.14)}$$

We can write the exponential part of equation (B.14) as

$$\exp\left(i\left[\mathbf{k}\cdot\mathbf{r}(jl)+\mathbf{k}'\cdot\mathbf{r}(j'l')\right]\right)$$

$$=\exp\left(i\left[\mathbf{k}+\mathbf{k}'\right]\cdot\mathbf{r}(jl)\right)\times\exp\left(i\mathbf{k}'\cdot\left[\mathbf{r}(j'l')-\mathbf{r}(jl)\right]\right) \qquad \text{(B.15)}$$

Equation (B.13) involves summation over both l and l'. We can modify this summation by defining:

$$\mathbf{r}(j'l')-\mathbf{r}(jl)=\mathbf{r}(j'l'')-\mathbf{r}(j0) \qquad \text{(B.16)}$$

and summing over l'' instead of l'. Thus equation (B.14) has the form:

$$\mathbf{u}^{\mathrm{T}}(jl)\cdot\mathbf{\Phi}\begin{pmatrix}jj'\\ ll''\end{pmatrix}\cdot\mathbf{u}(j'l'')$$

$$=\frac{1}{N\left(m_j m_{j'}\right)^{1/2}}\sum_{\mathbf{k},\nu}\sum_{\mathbf{k}',\nu'}\left\{\mathbf{e}^{\mathrm{T}}(j,\mathbf{k},\nu)\cdot\mathbf{\Phi}\begin{pmatrix}jj'\\ 0l''\end{pmatrix}\cdot\mathbf{e}(j',\mathbf{k}',\nu')\right.$$

$$\left.\times\exp\left(i\left[\mathbf{k}+\mathbf{k}'\right]\cdot\mathbf{r}(jl)\right)\exp\left(i\mathbf{k}'\cdot\left[\mathbf{r}(jl'')-\mathbf{r}(j0)\right]\right)Q(\mathbf{k},\nu)Q(\mathbf{k}',\nu')\right\} \text{(B.17)}$$

The summation over l in equation (B.13) implies, from equation (B.1) that $\mathbf{k}' = -\mathbf{k}$, and the factor of $1/N$ is also taken into account. Therefore the expression for the total potential energy, equation (B.13), becomes:

$$V=\frac{1}{2}\sum_{jj',l''}\frac{1}{\left(m_j m_{j'}\right)^{1/2}}\sum_{\mathbf{k},\nu,\nu'}\left\{\mathbf{e}^{\mathrm{T}}(j,\mathbf{k},\nu)\cdot\mathbf{\Phi}\begin{pmatrix}jj'\\ 0l''\end{pmatrix}\cdot\mathbf{e}(j',-\mathbf{k},\nu')\right.$$

$$\left.\times\exp\left(i\mathbf{k}\cdot\left[\mathbf{r}(j'l'')-\mathbf{r}(j0)\right]\right)Q(\mathbf{k},\nu)Q(-\mathbf{k},\nu')\right\}$$

$$=\sum_{\mathbf{k},\nu,\nu'}\mathbf{e}^{\mathrm{T}}(\mathbf{k},\nu)\cdot\mathbf{D}(\mathbf{k})\cdot\mathbf{e}(-\mathbf{k},\nu')\times Q(\mathbf{k},\nu)Q(-\mathbf{k},\nu') \qquad \text{(B.18)}$$

In the last step we have substituted in the *dynamical matrix* $\mathbf{D}(\mathbf{k})$, using equation (6.22). This has allowed us to use the matrix form of the eigenvectors $\mathbf{e}$ to include all the atoms. We can now use the orthogonality condition on the eigenvectors to note that terms with $\nu \neq \nu'$ must vanish. We also note that we are diagonalising the dynamical matrix, giving the normal eigenvalues:

$$\mathbf{e}^{\mathrm{T}}(\mathbf{k},\nu)\cdot\mathbf{D}(\mathbf{k})\cdot\mathbf{e}(-\mathbf{k},\nu') = \omega_\nu^2(\mathbf{k})\delta_{\nu,\nu'} \qquad \text{(B.19)}$$

Thus equation (B.18) ends up as

$$V = \frac{1}{2}\sum_{\mathbf{k},\nu}\omega_\nu^2(\mathbf{k})Q(\mathbf{k},\nu)Q(-\mathbf{k},\nu) \qquad \text{(B.20)}$$

which is the desired result.

Derivation of equation (A.5)

We start by noting the following Fourier transform:

$$\frac{2}{\sqrt{\pi}}\exp\left(-r^2\rho^2\right) = \frac{1}{4\pi^2}\int \rho^{-3}\exp\left(-k^2/4\rho^2\right)\exp(i\mathbf{k}\cdot\mathbf{r})\mathrm{d}\mathbf{k} \qquad \text{(B.21)}$$

We also note the following relationship derived in Squires (1978, App. A.4):

$$\sum_l \exp(i\mathbf{k}.\mathbf{r}(l)) = \frac{(2\pi)^3}{V_c}\sum_{\mathbf{G}}\delta(\mathbf{k}-\mathbf{G}) \qquad \text{(B.22)}$$

where V_c is the volume of the unit cell. We combine these two expressions by simply summing equation (B.18) over all values of l, but first noting that:

$$r_{ij}^2 = \left|\mathbf{r}(l) + \mathbf{r}_j - \mathbf{r}_i\right|^2 \qquad \text{(B.23)}$$

Thus we obtain:

$$\begin{aligned}
&\frac{2}{\sqrt{\pi}}\sum_l \exp\left(-r_{ij}^2(l)\rho^2\right) \\
&= \frac{1}{4\pi^2}\int \rho^{-3}\exp\left(-k^2/4\rho^2\right)\sum_l \exp(i\mathbf{k}\cdot\mathbf{r}(l))\exp\left(i\mathbf{k}\cdot\left(\mathbf{r}_j - \mathbf{r}_i\right)\right)\mathrm{d}\mathbf{k} \\
&= \frac{2\pi}{V_c}\int \rho^{-3}\exp\left(-k^2/4\rho^2\right)\sum_{\mathbf{G}}\delta(\mathbf{k}-\mathbf{G})\exp\left(i\mathbf{k}\cdot\left(\mathbf{r}_j - \mathbf{r}_i\right)\right)\mathrm{d}\mathbf{k} \\
&= \frac{2\pi}{V_c}\sum_{\mathbf{G}}\rho^{-3}\exp\left(-G^2/4\rho^2\right)\exp\left(iG\cdot\left(\mathbf{r}_j - \mathbf{r}_i\right)\right) \qquad \text{(B.24)}
\end{aligned}$$

Appendix C

Bose–Einstein distribution and the thermodynamic relations for phonons

The partition function for phonons is used to obtain the Bose–Einstein relation and the phonon free energy.

The Bose–Einstein distribution gives the mean number of phonons for any frequency at a given temperature T. To derive the distribution we need to start from the partition function,[1] Z, which is defined in its general form as

$$Z = \sum_{j=1}^{\infty} \exp\left(-E_j / k_B T\right) \tag{C.1}$$

where E_j is the energy of the j-th excited state. The partition function for the phonons associated with N normal modes of a crystal (not including the zero-point motion, see below) is therefore given as

$$Z = \sum_{n_1=0}^{\infty} \sum_{n_2=0}^{\infty} \cdots \sum_{n_N=0}^{\infty} \exp\left[-\frac{1}{k_B T}\left(n_1 \varepsilon_1 + n_2 \varepsilon_2 + \cdots + n_N \varepsilon_N\right)\right]$$

$$= \prod_k \left[\sum_{n_k=0}^{\infty} \exp\left(-n_k \varepsilon_k / k_B T\right)\right] \tag{C.2}$$

where the energy of an excitation, ε_k, is equal to $\hbar\omega_k$ for the branch and wave vector of the k-th normal mode of the crystal, and n_k is the number of phonons excited into the k-th normal mode. Equation (C.2) can be simplified by using the series result:

[1] The partition function has a central role in the development of statistical thermodynamics.

$$\sum_{n=0}^{\infty} \exp(-nx) = \frac{1}{1-\exp(-x)} \tag{C.3}$$

which enables us to write the partition function as

$$Z = \prod_k \frac{1}{1-\exp(-\varepsilon_k / k_B T)}$$
$$\Rightarrow \quad \ln Z = -\sum_k \ln\left[1-\exp(-\varepsilon_k / k_B T)\right] \tag{C.4}$$

The mean occupation number, n_k, of any state, ε_k, is obtained from the partition function by the standard result:

$$n_k = -k_B T \frac{\partial}{\partial \varepsilon_k} \ln Z$$
$$= \frac{1}{\exp(\varepsilon_k / k_B T)-1} = \frac{1}{\exp(\hbar\omega_k / k_B T)-1} \tag{C.5}$$

which is the Bose–Einstein distribution.

For the partition function in the complete case, we need to include the potential energy of the system, V, and the zero-point motion.[2] The only difference these make is that the partition function given by equations (C.2) and (C.4) needs to be multiplied by the factor:

$$\exp(-V / k_B T) \prod_k \exp(-\varepsilon_k / 2k_B T) \tag{C.6}$$

so that we need to add to the expression for $\ln Z$ the following term:

$$-\frac{1}{k_B T}\left(V + \frac{1}{2}\sum_k \varepsilon_k\right) \tag{C.7}$$

The free energy F is given as

$$F = -k_B T \ln Z$$
$$= V + \frac{1}{2}\sum_k \hbar\omega_k + k_B T \sum_k \ln\left[1-\exp\left(-\frac{\hbar\omega_k}{k_B T}\right)\right]$$
$$= V - \frac{1}{2}\sum_k \hbar\omega_k - k_B T \sum_k \ln n(\omega_k) \tag{C.8}$$

[2] We have not taken account of these terms until now in order to make the preceding equations less cumbersome.

To complete the full circle, we note that the expression for the internal energy E follows as

$$E = F + TS = F - T\frac{\partial F}{\partial T}$$

$$= \frac{k_B T^2}{Z}\frac{\partial Z}{\partial T}$$

$$= -\frac{1}{Z}\frac{\partial Z}{\partial \beta} \; ; \; \beta = \frac{1}{k_B T} \tag{C.9}$$

This result can be seen to be consistent with equation (C.1):

$$E = \frac{\sum_{j=1}^{\infty} E_j \exp\left(-E_j / k_B T\right)}{\sum_{j=1}^{\infty} \exp\left(-E_j / k_B T\right)} = -\frac{1}{Z}\frac{\partial Z}{\partial \beta} \tag{C.10}$$

The self-consistency of these results provides the justification of equation (C.8).

FURTHER READING

Born and Huang (1954) ch. 4, 16
Mandl (1971) pp 53–67, 153–154, 248–249

Appendix D

Landau theory of phase transitions

The basic ideas of Landau's theory of phase transition are reviewed, and the basic results for first- and second-order phase transitions are presented. The effects of coupling between the order parameter and other variables are described to an elementary level.

The order parameter

Central to any theory of phase transitions is the order parameter, which is conventionally denoted as Q (Salje 1990, p 13; Bruce and Cowley 1981, pp 13–40). The order parameter gives a quantitative measure of the extent to which the phase transition has changed the structure.[1] The order parameter concept is most easily understood for ferromagnetic phase transitions. We will take the simplest model of this transition, the *Ising model*, in which the moments on each site are constrained to lie only along the positive or negative z directions (Rao and Rao 1978, pp 175–183). In the high-temperature paramagnetic phase the atomic magnetic dipole moments do not have any preferential long-range alignment, so the moment on each site has an equal probability of pointing in either direction, giving an average moment of zero. Below the transition temperature the moments spontaneously align along one direction, although at non-zero temperatures thermal fluctuations mean that there is not complete order. The average moment on each site is therefore equal to the difference between the number of moments pointing along $+z$ and the number pointing along $-z$, scaled by the total number of sites in the crystal. The average moment (and hence the net macroscopic magnetisation) varies continuously from zero at the transition temperature to its maximum value at zero

[1] We assume the necessary condition that the symmetry of the low-temperature phase is a subgroup of the symmetry of the high-temperature phase.

temperature, and gives a quantitative measure of the degree of order in the system. The magnetisation can therefore be used as an order parameter for this transition. It is sometimes appropriate to scale the average magnetisation by the absolute size of the magnetic moment of a site in order to give the order parameter a value of unity at zero temperature, indicating complete order.

The order parameter for displacive phase transitions is not so easily defined. Given that a displacive transition follows from the condensation of a soft mode, we can take the amplitude of the distortion of the soft mode eigenvector below the transition temperature as an order parameter. For example, the angle of the rotation of the TiO_6 octahedra in $SrTiO_3$ can be taken to be the order parameter. For more complicated cases we can use the actual atomic displacements associated with the phase transition as measures of the order parameter. In these cases, however, we have a problem with the exact definition since distances are affected by thermal expansion and any spontaneous strains that accompany the transition. To some extent these problems can be removed by considering the changes in fractional coordinates, but in any case provided the displacements are not large these problems will not be crucial.

In the example of the ferromagnetic transition, we noted that the order parameter varies continuously from a value of zero at the transition temperature to a maximum value at zero temperature. Continuous transitions are common in the displacive case also, and are generally known, for historical reasons,[2] as *second-order* transitions. There are also transitions in which the order parameter jumps discontinuously to a non-zero value at the transition temperature; these are known as *first-order* phase transitions. The Landau theory provides a method of rationalising these different cases within a general framework in which the order parameter is the important quantity. It should be noted that Landau theory in its general use only has the purpose of providing a phenomenological description, which may not be quantitatively correct in detail. However, as discussed in Chapter 8, Landau theory actually works rather well for displacive phase transitions, and it is possible to give a physical interpretation of the various parameters.

[2] The order of a transition, first, second or whatever, was defined as the order of the differential of the free energy with respect to temperature that is discontinuous at the transition. For a first-order transition, the entropy is discontinuous at the transition, whereas for a second-order transition the heat capacity is discontinuous. Although higher-order cases may exist, it is conventional to use only the notions of first- and second-order transitions, and to note that a discontinuity in the order parameter indicates a discontinuity in the entropy.

Landau free energy for second-order phase transitions [3]

For a second-order phase transition we can expand the Gibbs free energy, G, as a polynomial in Q about the value for $Q = 0$, G_0:

$$G(Q) = G_0 + \frac{1}{2}aQ^2 + \frac{1}{4}bQ^4 + \cdots \quad \text{(D.1)}$$

Terms with odd powers of Q are usually, but not always, absent in the general case, in order to preserve the symmetry $G(-Q) = G(Q)$. We will meet cases where this does not hold below, but in such cases the phase transition cannot be second order. The equilibrium value of Q is that for which G is a minimum, which is expressed as the conditions:

$$\frac{\partial G}{\partial Q} = 0 \; ; \; \frac{\partial^2 G}{\partial Q^2} > 0 \quad \text{(D.2)}$$

When the coefficients a and b are both positive, $G(Q)$ has a single minimum at $Q = 0$, so the free energy expansion describes the system in the high-symmetry phase as its equilibrium state, which is generally the high-temperature phase. On the other hand, when a is negative (b still positive) $G(Q)$ has a maximum value at $Q = 0$ and minima at non-zero values of $\pm Q$, so that the equilibrium state of the system is the low-symmetry phase. Since the sign of the coefficient a determines which phase is stable, we can assume that a changes sign on cooling through the transition temperature T_c. The simplest form of a that has this property is

$$a = \tilde{a}(T - T_c) \quad \text{(D.3)}$$

where $\tilde{a}$ is a positive constant. It is assumed that the coefficient b (and all higher-order coefficients) have a much weaker temperature dependence such that they can be considered to be approximately constant. This final form of the free energy is known as the *Landau free energy*, G_L, and it is usually written without the constant term G_0:

$$G_L(Q) = \frac{1}{2}\tilde{a}(T - T_c)Q^2 + \frac{1}{4}bQ^4 \quad \text{(D.4)}$$

We have, for the moment, neglected terms higher than fourth order, since for many cases they are much smaller and of little effect.

A simple way of thinking about the Landau free energy is to relate it to the general thermodynamic relation:

[3] Salje 1990, ch 3; Bruce and Cowley 1981, Part I.

$$G_L(Q) = \Delta H - T\Delta S \tag{D.5}$$

where ΔH is the excess enthalpy (with respect to the high-symmetry phase), and ΔS is the excess entropy. The excess entropy is then given as

$$\Delta S = -\frac{\partial G_L}{\partial T} = -\frac{\tilde{a}}{2}Q^2 \tag{D.6}$$

and, by substituting the entropy back into equation (D.4), we have:

$$\Delta H = -\frac{1}{2}\tilde{a}T_c Q^2 + \frac{1}{4}bQ^4 \tag{D.7}$$

The excess enthalpy is thus a double-well function, with minima at

$$Q = \pm\left(\frac{\tilde{a}T_c}{b}\right)^{1/2} \tag{D.8}$$

Most of this enthalpy will be potential energy, but part of it may come from the zero-point phonon free energy.

Application of the equilibrium condition (D.2) gives the temperature dependence of the order parameter:

$$Q(T) = \left(\frac{\tilde{a}}{b}\right)^{1/2}(T_c - T)^{1/2} \tag{D.9}$$

The susceptibility χ is defined as[4]

$$\chi^{-1} = \frac{\partial^2 G}{\partial Q^2} \tag{D.10}$$

which gives

$$\begin{aligned} \chi^{-1} &= \tilde{a}(T - T_c) \quad \text{for} \quad T > T_c \\ \chi^{-1} &= 2\tilde{a}(T_c - T) \quad \text{for} \quad T < T_c \end{aligned} \tag{D.11}$$

Thus $\chi \to \infty$ at $T = T_c$. χ^{-1} is equal to the soft mode frequency in many cases.

The excess heat capacity, ΔC, follows from equations (D.6) and (D.9) as

[4] Technically χ is defined as $\partial Q/\partial E$, where E is a field that generates a non-zero value of Q, which may or may not have a physical realisation. From thermodynamics E is defined as $E = \partial G/\partial Q$. Equation (D.10) follows from these two definitions.

$$\Delta C = T\left(\frac{\partial \Delta S}{\partial T}\right) = 0 \quad \text{for} \quad T > T_c$$

$$\Delta C = \frac{\tilde{a}^2 T}{2b} \quad \text{for} \quad T < T_c \qquad \text{(D.12)}$$

with a discontinuity of $\tilde{a}^2 T_c/2b$ at the transition temperature.

First-order phase transitions

The Landau free energy can be applied to first-order phase transitions when the quartic term is negative. We then need to include the next-higher term in the free energy expansion so that the series expansion of $G_L(Q)$ converges sensibly, giving:

$$G_L = \frac{1}{2}\tilde{a}(T - T_c)Q^2 - \frac{1}{4}bQ^4 + \frac{1}{6}cQ^6 \qquad \text{(D.13)}$$

$G_L(Q)$ has a single minimum at $Q = 0$ for $T > T_c + b^2/4\tilde{a}c$. Immediately below this temperature $G_L(Q)$ has three minima, at $Q = 0$ and at

$$Q = \pm\left\{\frac{b + \left[b^2 - 4\tilde{a}c(T - T_c)\right]^{1/2}}{2c}\right\}^{1/2} \qquad \text{(D.14)}$$

The minima all have equal values of G_L (i.e. zero) when $Q^2 = 3b/4c$, which occurs at the temperature T_0:

$$T_0 = T_c + \frac{3b^2}{16\tilde{a}c} \qquad \text{(D.15)}$$

T_0 is the actual equilibrium phase transition temperature, at which the order parameter jumps in value from 0 to $\pm(3b/4c)^{1/2}$. For $T < T_0$ the order parameter follows the form given by equation (D.14).

The change in enthalpy at $T = T_0$ is obtained by substituting the value of Q at T_0 into equation (D.13) and subtracting the entropy (D.6), giving:

$$\Delta H(T = T_0) = -\frac{3\tilde{a}bT_0}{8c} = -\frac{\tilde{a}T_0 Q^2}{2} \qquad \text{(D.16)}$$

This is equivalent to the latent heat of the transition.

The susceptibility for this case follows the relation:

$$\chi^{-1} = \tilde{a}(T - T_c) \quad \text{for} \quad T > T_0$$
$$\chi^{-1} = \tilde{a}(T - T_c) - 3bQ^2 + 5cQ^4 \quad \text{for} \quad T < T_0 \tag{D.17}$$

At T_0, χ^{-1} jumps in value from $3b^2/16c$ to $12b^2/16c$ on cooling.

Tricritical phase transitions

The special case $b = 0$ in the Landau free energy is for a transition that is balanced between first and second order, and which is called a *tricritical* phase transition. The Landau free energy in this case has the form:

$$G_L(Q) = \frac{1}{2}\tilde{a}(T - T_c)Q^2 + \frac{1}{6}cQ^6 \tag{D.18}$$

with the transition at $T = T_c$ as for a second-order phase transition. The various results that follow are

$$Q(T) = \left(\frac{\tilde{a}}{c}\right)^{1/4}(T_c - T)^{1/4} \tag{D.19}$$

$$\Delta C = \frac{1}{4}\left(\frac{\tilde{a}^3}{c}\right)^{1/2} T(T_c - T)^{-1/2} \tag{D.20}$$

$$\chi^{-1} = \tilde{a}(T - T_c) \quad \text{for} \quad T > T_c$$
$$\chi^{-1} = 4\tilde{a}(T_c - T) \quad \text{for} \quad T < T_c \tag{D.21}$$

Note that in this case the value of the heat capacity diverges at the transition temperature.

Interaction between the order parameter and other variables

In many cases, the order parameter will interact with other distortions in the crystal (Salje 1990, ch. 5; Bruce and Cowley 1981, pp 43–44). We will take the example of strain, ε. The resultant behaviour depends on the symmetries of Q and ε. If they have the same symmetry, the two quantities can couple bilinearly in the free energy, and it turns out that Q and ε will have the same temperature dependence. Otherwise, the next-lowest coupling in the free energy will be between ε and Q^2, which results in the relationship $\varepsilon \propto Q^2$. An example is $SrTiO_3$, which undergoes a displacive phase transition in which the soft mode has a zone boundary wave vector. The strain can be of only one sign (the a and

b cell parameters are reduced relative to c), and so the strain must be proportional to Q^2. Higher-order couplings give rise to more complicated behaviour, which we will not investigate here.

Bilinear coupling

The Landau free energy for bilinear coupling is given as

$$G_L = \frac{1}{2}aQ^2 + \frac{1}{4}bQ^4 + \frac{1}{2}\eta\varepsilon Q + \frac{1}{2}C_{el}\varepsilon^2 \quad (D.22)$$

where η is the coupling constant, and C_{el} is the elastic constant in the absence of the coupling (called the *bare* elastic constant). The equilibrium condition $\partial G_L/\partial\varepsilon = 0$ gives:

$$\frac{\partial G_L}{\partial \varepsilon} = 0 = \frac{1}{2}\eta Q + C_{el}\varepsilon$$

$$\Rightarrow \varepsilon = -\frac{\eta Q}{2C_{el}} \quad (D.23)$$

When this result is substituted into the free energy of equation (D.22) we obtain:

$$G_L = \frac{1}{2}\left(a - \frac{\eta^2}{4C_{el}}\right)Q^2 + \frac{1}{4}bQ^4 \quad (D.24)$$

The quadratic coefficient is then equal to

$$a - \frac{\eta^2}{4C_{el}} = \tilde{a}\left(T - \tilde{T}_c\right) \quad (D.25)$$

where we have a new transition temperature $\tilde{T}_c$:

$$\tilde{T}_c = T_c + \frac{\eta^2}{4\tilde{a}C_{el}} \quad (D.26)$$

Thus the effect of bilinear coupling is to increase the transition temperature. All other quantities follow as previously. The renormalised elastic constant that follows $(\partial^2 G/\partial\varepsilon^2)$ acts like the inverse susceptibility, and falls to zero at the new transition temperature.

Quadratic coupling

The Landau free energy for quadratic coupling is given as

$$G_{\mathrm{L}} = \frac{1}{2}aQ^2 + \frac{1}{4}bQ^4 + \frac{1}{2}\xi\varepsilon Q^2 + \frac{1}{2}C_{\mathrm{el}}\varepsilon^2 \tag{D.27}$$

where ξ is the coupling constant. The equilibrium condition gives:

$$\frac{\partial G_{\mathrm{L}}}{\partial \varepsilon} = 0 = \frac{1}{2}\xi Q^2 + C_{\mathrm{el}}\varepsilon$$
$$\Rightarrow \ \varepsilon = -\frac{\xi Q^2}{2C_{\mathrm{el}}} \tag{D.28}$$

Substituting equation (D.28) into equation (D.29) gives:

$$G_{\mathrm{L}} = \frac{1}{2}aQ^2 + \frac{1}{4}\left(b - \frac{\xi^2}{2C_{\mathrm{el}}}\right)Q^4 \tag{D.29}$$

The effect of the quadratic coupling is to reduce the size of the coefficient of the quartic term, such that if the coupling is sufficiently strong the quartic coefficient becomes negative and the transition changes from second order to first order. This is the case for many ferroelectric phase transitions.

The importance of time scales

When the order parameter couples to the strain, it becomes important to note that the different quantities have different dynamic behaviour (Bruce and Cowley 1981, pp 43–44). If the phase transition is studied using a slow or static probe the strain can change as quickly as the order parameter, and the susceptibility $\chi(\omega = 0)$ can be calculated in the normal manner. On the other hand, if a high-frequency probe is used, such as light scattering, the strains will not be able to respond as fast as the order parameter (which will move as a soft optic phonon), and the susceptibility will need to be calculated under the condition of constant strain. Thus below the phase transition the high-frequency susceptibility, $\chi(\omega = \infty)$, will not be equivalent to $\chi(\omega = 0)$.

Multicomponent order parameters

The description of Landau theory that we have described here has been appropriate to the case when the order parameter has only one component (known as a one-dimensional order parameter). In many cases there are two or three components to the order parameter (Bruce and Cowley 1981, pp 61–64). For example, the order parameter in the ferroelectric cubic–tetragonal phase transition in $BaTiO_3$ has three components, corresponding to the polarisation lying along each of the three directions of the cubic phase. The full Landau theory for such systems will include expansions for each component, and will also include terms in which different components couple. The allowed terms are usually determined by symmetry. However, it is often found that the different components are only non-zero in different domains of the low-symmetry phase, with each domain having only a single non-zero component. Thus the full Landau expansion often reduces to an expansion for a single-order parameter. In a similar way, there may be several different-order parameters for different distortions (which may or may not be of the same symmetry) which can couple in the Landau free energy (Salje 1990, ch. 10). The comments above apply in this case also, except that the resultant behaviour may be more complicated than a simple reduction to an expansion of the free energy in one order parameter only.

Landau free energy functions with cubic terms

The model considered so far in this appendix is appropriate for the case when $G(Q) = G(-Q)$. This is not always so. For example, consider a phase transition from a hexagonal to a monoclinic structure that involves a shear strain ε in the basal plane of the hexagonal structure. The monoclinic angle will probably be equal to $\beta = 120°+\varepsilon$. It is evident that the energy of this phase will be different from that with the monoclinic angle $\beta = 120°-\varepsilon$, which will be the phase with the opposite sign of the order parameter. In this case we must include the odd-order terms in the Landau free energy. We will consider the simplest case, with terms up to fourth order only,[5] giving

$$G_L = \frac{1}{2}\tilde{a}(T - T_c)Q^2 - \frac{1}{3}bQ^3 + \frac{1}{4}cQ^4 \tag{D.30}$$

We have used a negative sign for the cubic term ($b > 0$) in order that the free energy is lower for positive values of Q.

The free energy has a single minimum at $Q = 0$ for $T > T_c + b^2/4\tilde{a}c$. A second minimum with higher energy appears below this temperature with

[5] More complicated cases are usually too complicated to be useful to the experimentalist!

$$Q = \frac{b + \left[b^2 - 4\tilde{a}c(T - T_c)\right]^{1/2}}{2c} \tag{D.31}$$

At the temperature $T_0 = T_c + 2b^2/9\tilde{a}c$ the two minima have the same free energy ($G = 0$), whereas below this temperature the minimum with the non-zero value of Q has the lowest energy. Therefore the phase transition occurs at $T = T_0$, when the order parameter changes discontinuously to the value $Q = 2b/3c$. The minimum with a negative value of Q appears at $T = T_c$, with

$$Q = \frac{b - \left[b^2 - 4\tilde{a}c(T - T_c)\right]^{1/2}}{2c} \tag{D.32}$$

The latent heat for the transition follows as

$$\Delta H(T = T_0) = -\frac{2\tilde{a}b^2 T_0}{9c^2} = -\frac{\tilde{a}T_0 Q^2}{2} \tag{D.33}$$

Finally, the susceptibility follows as

$$\begin{aligned} \chi^{-1} &= \tilde{a}(T - T_c) \quad \text{for} \quad T > T_0 \\ \chi^{-1} &= \tilde{a}(T - T_c) - 2bQ + 3cQ^2 \quad \text{for} \quad T < T_0 \end{aligned} \tag{D.34}$$

Unlike the case of the first-order phase transition discussed earlier, there is no discontinuity in χ^{-1} at the transition temperature. Instead, χ^{-1} falls to a minimum value of $2b^2/9c$ at $T = T_0$.

Critique of Landau theory

Landau theory was originally developed as a simple analysis of the behaviour close to the transition temperature. Since then the theory has been applied to a wide range of systems over a wide range of temperatures. It is essential to appreciate that the theory has a number of shortcomings which qualify its use. These are:

1 Landau theory neglects spatial and dynamic fluctuations of the order parameter, which are particularly important close to the transition temperature (Bruce and Cowley 1981). These fluctuations, known as critical fluctuations, will lower the transition temperature and modify the temperature dependence of the various quantities calculated in this section. It is possible to add a term that accounts for spatial fluctuations, but this will not produce

the correct results.[6] The failure of Landau theory close to the transition temperature is associated with the failure of the mean-field approximation, which asserts that all regions of the system experience the same environment. Critical fluctuations are particularly important in systems with short-range interactions, such as magnetic materials. Some perovskites, e.g. $SrTiO_3$, show important critical fluctuations (Müller and Berlinger 1971; Riste et al. 1971).

2. The coefficients in Landau theory may not be constant over a wide range of temperatures. For order–disorder phase transitions, the higher-order coefficients are proportional to temperature. For these systems it needs to be recalled that the Landau free energy is an expansion of the free energy in the vicinity of the transition temperature, over small temperature intervals in which these coefficients vary more slowly than the coefficient of the quadratic term.
3. Since the Landau free energy is an expansion of the free energy for small values of Q it is not a good approximation for large Q. The main implication of this is that the Landau free energy does not give the correct thermodynamic behaviour close to $T = 0$ K. In particular, we expect that the derivatives of the order parameter and the free energy should fall to zero at $T = 0$ K. The Landau free energy does not predict this behaviour (Salje et al. 1991).

We demonstrate in Chapter 8, however, that these limitations are not very important for displacive phase transitions. In particular, the mean-field approximation is found to work rather well for displacive phase transitions and we find in Chapter 8 that for displacive phase transitions the coefficients of the Landau theory are not expected to be very dependent on temperature. Even when the Landau free energy is a crude approximation to the true free energy, there are features of the Landau free energy (such as the symmetry) that enable detailed analysis to be taken quite some way.

FURTHER READING

Blinc and Zeks (1974)
Bruce and Cowley (1981)
Kittel (1976) ch. 13
Landau and Lifshitz (1980)
Salje (1990) ch. 1–6, 10
Toledano and Toledano (1988)

[6] These terms do nevertheless provide useful insights, such as applications to incommensurate phase transitions (Salje 1990, ch. 11).

Appendix E

Classical theory of coherent neutron scattering

We derive the essential formalism for scattering theory using classical methods. The results are directly applicable to elastic and inelastic coherent neutron scattering. The results are extended to account for quantum effects, and are recast into the formalism of scattering cross sections.

General scattering formalism

Scattering from single particles and ensemble

We first consider the scattering of a radiation beam (electromagnetic or neutron) from a point particle. This is illustrated in Figure E.1 below. The particle is at position $\mathbf{r}$ with respect to the origin. $\mathbf{k}_i$ and $\mathbf{k}_f$ are the initial (incoming) and final (outgoing) wave vectors of the radiation beam respectively. When compared with scattering from the origin, the additional contributions to the path length are $l_i = r\cos\alpha$ and $l_f = r\cos\beta$. These give a net phase difference of:

$$\begin{aligned} l_i k_i - l_f k_f &= k_i r\cos\alpha - k_f r\cos\beta \\ &= \left(\mathbf{k}_i - \mathbf{k}_f\right)\cdot\mathbf{r} = \mathbf{Q}\cdot\mathbf{r} \end{aligned} \tag{E.1}$$

where $\mathbf{Q} = \mathbf{k}_i - \mathbf{k}_f$ is called the *scattering vector* or *wave vector transfer*.

This analysis is readily extended to coherent scattering from an ensemble of point particles, whether in an atom, crystal, liquid or whatever. The total scattering amplitude relative to some origin, $F(\mathbf{Q})$, is given as a sum over all the individual phase factors:

$$F(\mathbf{Q}) = \sum_j b_j \exp\left(i\mathbf{Q}\cdot\mathbf{r}_j\right) \tag{E.2}$$

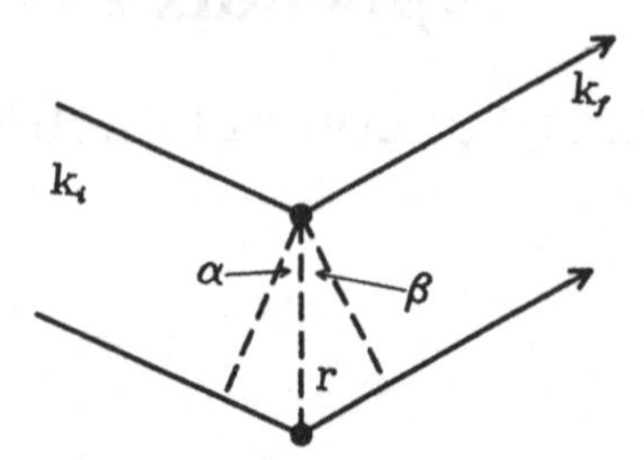

Figure E.1: Scattering diagram.

where b_j gives the contribution to the net scattering amplitude from the j-th atom. The intensity of the scattered beam is simply the square of the modulus of the amplitude:

$$I(\mathbf{Q}) = \left\langle \left| F(\mathbf{Q}) \right|^2 \right\rangle \tag{E.3}$$

The brackets $\langle \cdots \rangle$ denote an average over time; $F(\mathbf{Q})$ as written in equation (E.2) is an instantaneous amplitude, whereas any typical experiment is performed over a period of time that is long in comparison with the time scale of microscopic fluctuations.

The quantity b is, in the case of neutron scattering, called the *scattering length*, because it is common to define the intensity (which is proportional to b^2) as a cross-sectional area. For the case of thermal neutron scattering, b is independent of $\mathbf{Q}$, whereas for X-ray scattering b is strongly dependent on the magnitude of $\mathbf{Q}$ over the range of useful scattering angles.

Fourier analysis

We will now relate these results to a general Fourier analysis. The instantaneous particle density of the ensemble is given as a sum of delta functions, each of which represents a single point particle at position $\mathbf{r}_j$:

$$\rho(\mathbf{r}) = \sum_j b_j \delta\left(\mathbf{r} - \mathbf{r}_j\right) \tag{E.4}$$

We have included the scattering length of each particle in this expression in order that the density is equivalent to that seen by the incident radiation. The delta functions used in equation (E.4) have the following usual properties:

$$\delta(\mathbf{r}-\mathbf{r}_j) = 0 \quad \text{for } \mathbf{r} \neq \mathbf{r}_j$$
$$= \infty \quad \text{for } \mathbf{r} = \mathbf{r}_j \tag{E.5}$$

$$\int \delta(\mathbf{r}-\mathbf{r}_j)\mathrm{d}\mathbf{r} = 1 \tag{E.6}$$

The Fourier transform of the delta function is readily obtained by performing the integral over the two distinct ranges $\mathbf{r} \neq \mathbf{r}_j$ and $\mathbf{r} = \mathbf{r}_j$:

$$\int \delta(\mathbf{r}-\mathbf{r}_j)\exp(\mathrm{i}\mathbf{Q}\cdot\mathbf{r})\mathrm{d}\mathbf{r}$$
$$= \int_{\mathbf{r}\neq\mathbf{r}_j} \delta(\mathbf{r}-\mathbf{r}_j)\exp(\mathrm{i}\mathbf{Q}\cdot\mathbf{r})\mathrm{d}\mathbf{r} + \int_{\mathbf{r}=\mathbf{r}_j} \delta(\mathbf{r}-\mathbf{r}_j)\exp(\mathrm{i}\mathbf{Q}\cdot\mathbf{r})\mathrm{d}\mathbf{r} \tag{E.7}$$

The first integral on the right hand side is equal to zero, because the delta function is zero over the whole range $\mathbf{r} \neq \mathbf{r}_j$. The second integral on the right hand side becomes:

$$\exp(\mathrm{i}\mathbf{Q}\cdot\mathbf{r}_j)\int_{\mathbf{r}=\mathbf{r}_j} \delta(\mathbf{r}-\mathbf{r}_j)\mathrm{d}\mathbf{r} = \exp(\mathrm{i}\mathbf{Q}\cdot\mathbf{r}_j) \tag{E.8}$$

Thus the Fourier transform of the particle density (E.4) is readily obtained:

$$\rho(\mathbf{Q}) = \int \rho(\mathbf{r})\exp(i\mathbf{Q}\cdot\mathbf{r})\mathrm{d}\mathbf{r} = \sum_j b_j \int \delta(\mathbf{r}-\mathbf{r}_j)\exp(i\mathbf{Q}\cdot\mathbf{r})\mathrm{d}\mathbf{r}$$
$$= \sum_j b_j \exp(i\mathbf{Q}\cdot\mathbf{r}_j) \tag{E.9}$$

This is equal to the scattering amplitude, $F(\mathbf{Q})$, as given by equation (E.2). Thus the scattered intensity is simply given as

$$I(\mathbf{Q}) = \langle \rho(\mathbf{Q})\rho(-\mathbf{Q})\rangle = \sum_{j,k} b_j b_k \left\langle \exp\left(i\mathbf{Q}\cdot\left[\mathbf{r}_j - \mathbf{r}_k\right]\right)\right\rangle \tag{E.10}$$

Bragg and diffuse scattering

One important point about the intensity of the scattered beam is that it contains information only about relative positions of pairs of particles, and essentially no information about absolute positions. Let us consider this point by making contact with Bragg scattering. The Bragg intensity is given by:

$$I_{\text{Bragg}}(\mathbf{Q}) = \left|\langle \rho(\mathbf{Q})\rangle\right|^2 \tag{E.11}$$

where $\langle \rho(\mathbf{Q}) \rangle$ is equivalent to the crystallographic structure factor. The equation for the Bragg intensity is not the same as the scattering intensity given by equation (E.10). The residual intensity, namely the total scattering minus the Bragg scattering, is called the *diffuse scattering*:

$$I_{\text{Diffuse}}(\mathbf{Q}) = \langle \rho(\mathbf{Q})\rho(-\mathbf{Q}) \rangle - \left| \langle \rho(\mathbf{Q}) \rangle \right|^2 \tag{E.12}$$

It is a standard result from crystallography that $\langle \rho(\mathbf{Q}) \rangle$ is zero unless $\mathbf{Q}$ is a reciprocal lattice vector, whereas the diffuse intensity can be non-zero for all values of $\mathbf{Q}$. We can explore equation (E.12) in more detail. We define the time-dependent position of a particle $\mathbf{r}_j(t)$ as:

$$\mathbf{r}_j(t) = \mathbf{R}_j + \mathbf{u}_j(t) \tag{E.13}$$

where $\mathbf{u}_j(t)$ gives the instantaneous displacement of the particle from its equilibrium position $\mathbf{R}_j$. Thus we can rewrite our Fourier expressions:

$$\rho(\mathbf{Q},t) = \sum_j b_j \exp\left(i\mathbf{Q} \cdot \left[\mathbf{R}_j + \mathbf{u}_j(t)\right]\right) \tag{E.14}$$

$$\langle \rho(\mathbf{Q}) \rangle = \sum_j b_j \left\langle \exp\left(i\mathbf{Q} \cdot \left[\mathbf{R}_j + \mathbf{u}_j\right]\right) \right\rangle = \sum_j b_j \left\langle \exp\left(i\mathbf{Q} \cdot \mathbf{u}_j\right) \right\rangle \exp\left(i\mathbf{Q} \cdot \mathbf{R}_j\right) \tag{E.15}$$

For harmonic motions the average over the fluctuating part can be expressed in the form of the crystallographic *temperature* (or *Debye–Waller*) factor:

$$\left\langle \exp\left(i\mathbf{Q} \cdot \mathbf{u}_j\right) \right\rangle = \exp\left(-W_j(\mathbf{Q})\right) \tag{E.16}$$

where

$$W_j(\mathbf{Q}) = \frac{1}{2} \left\langle \left| \mathbf{Q} \cdot \mathbf{u}_j \right|^2 \right\rangle \tag{E.17}$$

Hence,

$$\langle \rho(\mathbf{Q}) \rangle = \sum_j b_j \exp\left(-W_j(\mathbf{Q})\right) \exp\left(i\mathbf{Q} \cdot \mathbf{R}_j\right) \tag{E.18}$$

and

$$I_{\text{Bragg}}(\mathbf{Q}) = \sum_{j,k} b_j b_k \exp\left(-\left[W_j(\mathbf{Q}) + W_k(\mathbf{Q})\right]\right)\exp\left(i\mathbf{Q}\cdot\left[\mathbf{R}_j - \mathbf{R}_k\right]\right)$$

$$= \left|\sum_j b_j \exp\left(-W_j(\mathbf{Q})\right)\exp\left(i\mathbf{Q}\cdot\mathbf{R}_j\right)\right|^2 \qquad \text{(E.19)}$$

This gives information only about the mean relative positions. If we take the Fourier transform of the Bragg intensity, we will end up with a function that involves delta functions with poles at ($\mathbf{R}_j$–$\mathbf{R}_k$) that are convoluted with the size of the particles (as given by the scattering factor $b_j(\mathbf{Q})$) and the temperature factors. The temperature factors are Gaussian in **Q**, and hence their transforms are also Gaussian in **u**. But aside from the details of the convolutions, the Fourier transform of the Bragg intensity contains strong maxima at the positions ($\mathbf{R}_j$–$\mathbf{R}_k$) that are weighted by the product $b_j b_k$. This can be recognised as the famous *Patterson function*, which is often viewed simply as a mathematical construction for the aid of crystallographers but which is in fact the fundamental quantity in scattering theory.

Time dependence and the inelastic scattering function

The above analysis has neglected all time dependence, assuming that there is no change in energy of the scattered beam and that all particles scatter at the same time. We now allow for a change in energy $\hbar\omega$ between the initial (i) and final (f) beams, such that:

$$\omega = \omega_i - \omega_f \qquad \text{(E.20)}$$

where $\hbar\omega_i$ is the energy of the incoming neutron beam and, $\hbar\omega_f$ is the energy of the outgoing beam.

For scattering from a particle at time t compared with the time origin, the scattered amplitude has an additional phase factor of $\exp(-i\omega t)$. Thus the contribution to the scattered intensity for any time t is equal to the time-dependent density operator:

$$\rho(\mathbf{Q},t) = \sum_j b_j \exp\left(i\mathbf{Q}\cdot\mathbf{r}_j(t)\right)\exp(-i\omega t) \qquad \text{(E.21)}$$

But just as the total scattered intensity involves a sum over all atoms, we need to extend the above sum to include all times. Thus the density operator becomes:

$$\rho(\mathbf{Q},\omega)=\int\sum_j b_j \exp\left(i\mathbf{Q}\cdot\mathbf{r}_j(t)\right)\exp(-i\omega t)\mathrm{d}t \tag{E.22}$$

The resultant scattered intensity then follows from equation (E.10) as:

$$\begin{aligned} I(\mathbf{Q},\omega) &= \int \rho(\mathbf{Q},t')\exp(-i\omega t')\mathrm{d}t' \times \int \rho(-\mathbf{Q},t'')\exp(-i\omega t'')\mathrm{d}t'' \\ &= \iint \rho(\mathbf{Q},t')\rho(-\mathbf{Q},t'+t)\exp(-i\omega t)\mathrm{d}t'\mathrm{d}t \end{aligned} \tag{E.23}$$

where we have used the property that a product of two Fourier transforms is equal to the Fourier transform of a convolution function. The integral over t' essentially gives a thermal average, so that we conclude that the dynamic scattering intensity is given as

$$I(\mathbf{Q},\omega) \propto S(\mathbf{Q},\omega) = \int \langle \rho(\mathbf{Q},0)\rho(-\mathbf{Q},t)\rangle \exp(-i\omega t)\mathrm{d}t \tag{E.24}$$

The quantity $S(\mathbf{Q}, \omega)$ is called the *scattering function*. We can simply write equation (E.24) as

$$S(\mathbf{Q},\omega) = \int F(\mathbf{Q},t)\exp(-i\omega t)\mathrm{d}t \tag{E.25}$$

where $F(\mathbf{Q}, t)$ is called the *intermediate scattering function*, and is therefore defined as

$$F(\mathbf{Q},t) = \langle \rho(\mathbf{Q},0)\rho(-\mathbf{Q},t)\rangle \tag{E.26}$$

We can view the intermediate scattering function as a correlation function (see Appendix F), which gives information about the dynamic fluctuations of the density of the crystal. These fluctuations will be dependent on the wave vector, and the intermediate scattering function will give the fluctuations for each wave vector. $S(\mathbf{Q}, \omega)$ gives the corresponding power spectrum.

We note that the integral of $S(\mathbf{Q}, \omega)$ over all frequencies gives equation (E.10):

$$\int S(\mathbf{Q},\omega)\mathrm{d}\omega = F(\mathbf{Q},0) = \langle \rho(\mathbf{Q})\rho(-\mathbf{Q})\rangle = S(\mathbf{Q}) \tag{E.27}$$

where we have performed an inverse Fourier transform on $F(\mathbf{Q}, t)$ and set $t = 0$. Thus diffuse scattering from the present theory is given as a sum over the scattering from all fluctuations (lattice vibrations).[1]

[1] There will of course be other contributions to diffuse scattering, such as from defects, but for many systems this is the dominant source of diffuse scattering. As it increases with temperature, it is often given the term *thermal diffuse scattering* to differentiate it from other sources of diffuse scattering.

The basic results of this section do not change in the full quantum-mechanical treatment of scattering theory. However, there will be a difference between the classical and quantum results when the functions derived in this appendix are evaluated. The principal difference between the classical and quantum versions of scattering theory is that the classical model gives time-reversal properties that are inconsistent with the quantum picture. Squires (1978, pp 65–70) gives the essential quantum-mechanical relationships that are not obeyed by classical mechanics:

$$F(\mathbf{Q},t) = F^*(\mathbf{Q},-t) \tag{E.28}$$

$$F(\mathbf{Q},t) = F\left(-\mathbf{Q},-t + i\hbar / k_\mathrm{B}T\right) \tag{E.29}$$

$$S(\mathbf{Q},\omega) = S^*(\mathbf{Q},\omega) \tag{E.30}$$

$$S(\mathbf{Q},\omega) = \exp\left(\hbar\omega / k_\mathrm{B}T\right)S(-\mathbf{Q},-\omega) \tag{E.31}$$

For a classical system these functions should be independent of the sign of $\mathbf{Q}$ or ω. The origin of the relationships expressed by equation (E.28)–(E.31) is the principle of detailed balance. For scattering from phonons, there will be a difference between creation and absorption of phonons, in that for absorption the phonons need to be thermally excited. Thus at low temperatures there will be few phonons for absorption but phonon creation will still be possible. This accounts for the role of temperature in these relationships.

Scattering cross section

In practice the quantity that is measured in any scattering experiment is a cross section.[2] For measurements that simply detect all the neutrons that are scattered into the solid angle $\mathrm{d}\Omega$ in the direction given by the polar angles θ and ϕ, we can define the *differential cross section* as

$$\frac{\mathrm{d}\sigma}{\mathrm{d}\Omega} = \frac{n}{J\mathrm{d}\Omega} \tag{E.32}$$

where n is the number of neutrons scattered per second into $\mathrm{d}\Omega$ in the direction (θ, ϕ), and J is the incident neutron flux.

When the experiment is set up to analyse the energy of the scattered neutrons, the quantity measured is the *partial* or *double differential cross section*:

[2] In a quantum treatment of neutron scattering the theory is developed from the beginning in terms of the cross section (Placzek and Van Hove 1954; Van Hove 1954), as for example in Squires (1978, ch. 1–4), Marshall and Lovesey (1971) and Lovesey (1984).

$$\frac{d^2\sigma}{d\Omega dE} = \frac{n'}{J d\Omega dE} \tag{E.33}$$

where n' is the number of neutrons scattered per second into $d\Omega$ in the direction (θ, ϕ) with final energy between E and $E + dE$. These two cross sections are clearly related by

$$\frac{d\sigma}{d\Omega} = \int_0^\infty \frac{d^2\sigma}{d\Omega dE} dE \tag{E.34}$$

The total cross section σ_{tot} is then given as

$$\sigma_{tot} = \int \frac{d\sigma}{d\Omega} d\Omega \tag{E.35}$$

It can be shown from a more complete derivation of the neutron scattering function (e.g. Squires 1978, ch. 4) that the cross section and the inelastic scattering factor given by equation (E.24) are simply related by

$$\frac{d^2\sigma}{d\Omega dE} = \frac{k_f}{k_i} S(\mathbf{Q}, \omega) \tag{E.36}$$

FURTHER READING

Marshall and Lovesey (1971)
Squires (1978)

Appendix F
Time correlation functions

Any experiment will give information that is related to a static or dynamic correlation function, and we have already shown that correlation functions are extremely useful for neutron scattering and molecular dynamics simulations. The main aspects of correlation functions are described in this appendix.

Time-dependent correlation functions

Consider a time-dependent quantity, $x(t)$, which may be a particle coordinate, velocity or other quantity. Typically $x(t)$ will oscillate about a mean value, but its detailed time dependence will hide so many independent influences that it will have the appearance of being a random oscillation. Of course, $x(t)$ will not really vary randomly, but will contain important information about the system. The question is then how to extract any information from the function $x(t)$. All the influences that perturb the behaviour of $x(t)$ mean that we will never be able to predict the actual behaviour of $x(t)$, nor probably will we ever want to! Instead we will need to consider trends within the behaviour of $x(t)$, and this is the basic philosophy behind the *correlation function*.

The correlation function is a tool that enables us to describe the average way the quantity x will change with time. The simplest correlation function for a quantity with a mean value of zero, $C(t)$, is defined as

$$C(t) = \frac{\langle x(0)x(t) \rangle}{\langle |x(0)|^2 \rangle} \tag{F.1}$$

where the brackets $\langle \cdots \rangle$ denote an average over all starting times.[1] Thus we can rewrite equation (F.1) as

[1] The correlation function for which the two variables are the same is called an *autocorrelation function*.

$$C(t) = \frac{(\lim \mathcal{T} \to \infty)\dfrac{1}{\mathcal{T}}\displaystyle\int_0^{\mathcal{T}} x(t')x(t+t')\mathrm{d}t'}{\langle x^2 \rangle} \tag{F.2}$$

If we have a number of identical particles in the system, the brackets will also include the average over all particles. In general $C(t)$ will be defined to have a value of unity at $t = 0$, as in equations (F.1) and (F.2), although this is not always necessary. Normalisation has the advantage that it highlights the statistical nature of the correlation function. When $C(t) = 1$, as at $t = 0$, there is a complete correlation of values, with a 100% certainty that the quantity will have exactly the same value at two times separated by a time interval t. When $C(t) = 0$, as it may often do when t is very large, there is no relationship between any two values of the quantity over this particular time interval. Thus the correlation function tells you how likely it is that values of x at two times separated by a period of time t are related

Consider now two examples. In the first $x(t)$ varies as $\sin(\omega t)$. The correlation function, normalised to have a value of unity at $t = 0$, then has the form:

$$C(t) = \frac{\omega}{\pi}\int_0^{2\pi/\omega} \sin(\omega t')\sin(\omega t + \omega t')\mathrm{d}t' = \cos(\omega t) \tag{F.3}$$

In this case the correlation function exactly reflects the sinusoidal nature of the original function. Our second example is for the case when x can have only two equally probable values, ± 1, and the probability of x changing its value during an infinitesimal time interval $\mathrm{d}t$ is $\mathrm{d}t/\tau$. The corresponding correlation function is then equal to

$$C(t) = \exp(-|t|/\tau) \tag{F.4}$$

The time constant τ gives the average length of time between changes of the value of x. In practice many correlation functions lie between the two extremes of equations (F.3) and (F.4). Moreover, in general equation (F.4) will need to be modified to take account of the time taken for x to change its value, a factor that will be important for small values of time.

Power spectra

Most time-dependent quantities are determined by the superposition of a large number of vibrations, so we are less interested in the time dependence itself than in the distribution of vibrational frequencies that contribute to the behaviour of our quantity. Thus we are interested in the *power spectrum* of $C(t)$, which is given by the Fourier transform $Z(\omega)$:

$$Z(\omega) = \int C(t)\exp(-i\omega t)\mathrm{d}t \tag{F.5}$$

The power spectrum of equation (F.3) is a delta function at a frequency of ω, showing that the behaviour of $x(t)$ is determined by a single vibrational mode. The power spectrum of equation (F.4) is a Lorentzian centred about zero frequency:

$$Z(\omega) = \int_{-\infty}^{\infty} \exp(-|t|/\tau)\exp(-i\omega t)\mathrm{d}t = \frac{2\tau}{1+(\omega\tau)^2} \tag{F.6}$$

A typical exponentially-damped vibration, such as a normal mode vibration that is damped by anharmonic effects, will have a correlation function that is the product of a cosine and exponential function. The corresponding power spectrum will therefore be the convolution of a delta function at the vibrational frequency with the Lorentzian of equation (F.6), resulting in a Lorentzian centred about the vibrational frequency.

The power spectrum of a correlation function can be obtained without having to construct the correlation function, using a result known as the the *Wiener–Khintchine* theorem. If we define $a(\omega)$ as the Fourier transform of our initial time-dependent quantity $x(t)$:

$$a(\omega) = \int x(t)\exp(-i\omega t)\mathrm{d}t \tag{F.7}$$

the power spectrum, $Z(\omega)$, and correlation function, $C(t)$, are given as

$$Z(\omega) = |a(\omega)|^2 \tag{F.8}$$

$$C(t) = \int Z(\omega)\exp(i\omega t)\mathrm{d}t \tag{F.9}$$

The results of this section are particularly useful for the analysis of data from molecular dynamics simulations (Chapter 12), for which analyses of correlation functions and the associated power spectra are often the major part of an investigation.

Example: the velocity autocorrelation function and the phonon density of states

One important correlation function, which serves as a good summary example, is for the velocity of an atom in a harmonic crystal. The velocity of the j-th atom in the l-th unit cell, $\dot{\mathbf{u}}(jl, t)$, is given as

$$\dot{\mathbf{u}}(jl,t) = \frac{-i}{\left(Nm_j\right)^{1/2}} \sum_{\mathbf{k},\nu} \omega(\mathbf{k},\nu)\mathbf{e}(j,\mathbf{k},\nu)\exp\left(i\mathbf{k}\cdot\mathbf{r}(jl)\right)Q(\mathbf{k},\nu,t) \quad \text{(F.10)}$$

where the symbols are defined in Chapter 4. The time dependence on the right hand side of equation (F.10) is contained in the normal mode coordinate $Q(\mathbf{k}, \nu)$. The important quantity for the correlation function before performing the averaging is $\left|\dot{\mathbf{u}}(jl,t)\cdot\dot{\mathbf{u}}(jl,0)\right|$. If we multiply by m_j and add all the atoms in the unit cell, we obtain the quantity:

$$\sum_j m_j \left\langle \left|\dot{\mathbf{u}}(jl,t)\cdot\dot{\mathbf{u}}(jl,0)\right|\right\rangle = \frac{1}{N}\sum_{\mathbf{k},\nu} \omega^2(\mathbf{k},\nu)\left\langle Q(\mathbf{k},\nu,t)Q(-\mathbf{k},\nu,0)\right\rangle \quad \text{(F.11)}$$

where we have also averaged over all unit cells. The correlation function on the right hand side of equation (F.11) can be written as

$$\left\langle Q(\mathbf{k},\nu,t)Q(-\mathbf{k},\nu,0)\right\rangle = \left\langle \left|Q(\mathbf{k},\nu)\right|^2\right\rangle \cos\left(\omega(\mathbf{k},\nu)t\right) \quad \text{(F.12)}$$

which in the classical limit is equal to

$$\left\langle Q(\mathbf{k},\nu,t)Q(-\mathbf{k},\nu,0)\right\rangle = \frac{k_{\mathrm{B}}T}{\omega^2(\mathbf{k},\nu)}\cos\left(\omega(\mathbf{k},\nu)t\right) \quad \text{(F.13)}$$

This leads to the final classical result:

$$\sum_j m_j \left\langle \left|\dot{\mathbf{u}}(jl,t)\cdot\dot{\mathbf{u}}(jl,0)\right|\right\rangle = \frac{k_{\mathrm{B}}T}{N}\sum_{\mathbf{k},\nu}\cos\left(\omega(\mathbf{k},\nu)t\right) \quad \text{(F.14)}$$

It can be seen from this result that the Fourier transform will give an equal-weight contribution for each normal mode. Thus the power spectrum of the mass-weighted velocity correlation function is equal to the phonon density of states.

Appendix G
Commutation relations

The main commutation relations used in Chapter 11 are derived in this appendix.

The commutation relation for two operators, $\hat{a}$ and $\hat{b}$, is defined as

$$\left[\hat{a},\hat{b}\right] = \hat{a}\hat{b} - \hat{b}\hat{a} \tag{G.1}$$

The commutation relation for the two fundamental operators, the position $\hat{q}$, and the momentum $\hat{p}$, is given as

$$\hat{q} = q \ ; \ \hat{p} = -i\hbar\frac{\partial}{\partial q}$$
$$\left[\hat{q},\hat{p}\right] = i\hbar \tag{G.2}$$

We can define operators for the normal mode coordinates, following equation (4.2):

$$\hat{\mathbf{q}}(jl,t) = \frac{1}{\left(Nm_j\right)^{1/2}}\sum_{\mathbf{k},\nu}\mathbf{e}(j,\mathbf{k},\nu)\exp\left(i\mathbf{k}\cdot\mathbf{r}(jl)\right)\hat{Q}(\mathbf{k},\nu) \tag{G.3}$$

$$\hat{\mathbf{p}}(jl,t) = \frac{1}{\left(Nm_j\right)^{1/2}}\sum_{\mathbf{k},\nu}\mathbf{e}(j,\mathbf{k},\nu)\exp\left(i\mathbf{k}\cdot\mathbf{r}(jl)\right)\hat{P}(\mathbf{k},\nu) \tag{G.4}$$

The normal mode operators are therefore defined as

$$\hat{Q}(\mathbf{k},\nu)=\frac{1}{\sqrt{N}}\sum_{j,l}m_j^{1/2}\exp(-i\mathbf{k}\cdot\mathbf{r}(jl))\mathbf{e}^*(j,\mathbf{k},\nu)\cdot\hat{\mathbf{q}}(j,l) \tag{G.5}$$

$$\hat{P}(\mathbf{k},\nu)=\frac{1}{\sqrt{N}}\sum_{j,l}m_j^{-1/2}\exp(-i\mathbf{k}\cdot\mathbf{r}(jl))\mathbf{e}^*(j,\mathbf{k},\nu)\cdot\hat{\mathbf{p}}(j,l) \tag{G.6}$$

The commutation relations for the normal mode operators follow:

$$\begin{aligned}\left[\hat{Q}(\mathbf{k},\nu),\hat{P}(-\mathbf{k},\nu)\right]&=\sum_{j,l}\sum_{j',l'}\left\{\sqrt{\frac{m_j}{m_{j'}}}\exp\left(i\mathbf{k}\cdot\left(\mathbf{r}(j'l')-\mathbf{r}(jl)\right)\right)\right.\\&\left.\times\sum_{\alpha,\beta}e^*_\alpha(j,\mathbf{k},\nu)e^*_\beta(j',-\mathbf{k},\nu)\left[\hat{q}_\alpha(jl),\hat{p}_\beta(j'l')\right]\right\}\\&=\frac{1}{N}\sum_{j,l,l'}\exp\left(i\mathbf{k}\cdot\left(\mathbf{r}(jl')-\mathbf{r}(jl)\right)\right)\mathbf{e}^*(j,\mathbf{k},\nu)\cdot\mathbf{e}^*(j,-\mathbf{k},\nu)i\hbar\\&=\frac{i\hbar}{N}\sum_{l,l'}\exp\left(i\mathbf{k}\cdot\left(\mathbf{r}(l')-\mathbf{r}(l)\right)\right)=i\hbar\end{aligned} \tag{G.7}$$

Equation (G.7) is readily generalised:

$$\left[\hat{Q}(\mathbf{k},\nu),\hat{P}(\mathbf{k}',\nu')\right]=i\hbar\delta_{\mathbf{k},-\mathbf{k}'}\delta_{\nu,\nu'} \tag{G.8}$$

We next form the products of the creation and annihilation operators:

$$\begin{aligned}\hat{a}^+(\mathbf{k},\nu)\hat{a}(\mathbf{k},\nu)&=\frac{1}{2\hbar\omega(\mathbf{k},\nu)}\left\{\omega^2(\mathbf{k},\nu)\hat{Q}^+(\mathbf{k},\nu)\hat{Q}(\mathbf{k},\nu)\right.\\&\left.+\hat{P}^+(\mathbf{k},\nu)\hat{P}(\mathbf{k},\nu)+i\omega(\mathbf{k},\nu)\left(\hat{Q}^+(\mathbf{k},\nu)\hat{P}(\mathbf{k},\nu)-\hat{P}^+(\mathbf{k},\nu)\hat{Q}(\mathbf{k},\nu)\right)\right\}\end{aligned} \tag{G.9a}$$

$$\begin{aligned}\hat{a}(\mathbf{k},\nu)\hat{a}^+(\mathbf{k},\nu)&=\frac{1}{2\hbar\omega(\mathbf{k},\nu)}\left\{\omega^2(\mathbf{k},\nu)\hat{Q}^+(\mathbf{k},\nu)\hat{Q}(\mathbf{k},\nu)\right.\\&\left.+\hat{P}^+(\mathbf{k},\nu)\hat{P}(\mathbf{k},\nu)-i\omega(\mathbf{k},\nu)\left(\hat{Q}(\mathbf{k},\nu)\hat{P}^+(\mathbf{k},\nu)-\hat{P}(\mathbf{k},\nu)\hat{Q}^+(\mathbf{k},\nu)\right)\right\}\end{aligned} \tag{G.9b}$$

Hence we obtain the commutation relations:

$$\left[\hat{a}(\mathbf{k},\nu),\hat{a}^{+}(\mathbf{k},\nu)\right]=\frac{1}{2\hbar\omega(\mathbf{k},\nu)}\left\{i\omega(\mathbf{k},\nu)\left[\hat{P}(\mathbf{k},\nu),\hat{Q}^{+}(\mathbf{k},\nu)\right]\right.$$

$$\left.+i\omega(\mathbf{k},\nu)\left[\hat{P}^{+}(\mathbf{k},\nu),\hat{Q}(\mathbf{k},\nu)\right]\right\}=1 \qquad \text{(G.10a)}$$

$$\left[\hat{a}^{+}(\mathbf{k},\nu),\hat{a}(\mathbf{k},\nu)\right]=-1 \qquad \text{(G.10b)}$$

$$\left[\hat{a}(\mathbf{k},\nu),\hat{a}(\mathbf{k},\nu)\right]=\left[\hat{a}^{+}(\mathbf{k},\nu),\hat{a}^{+}(\mathbf{k},\nu)\right]=0 \qquad \text{(G.10c)}$$

We next consider how the creation and annihilation operators commute with the Hamiltonian:

$$\left[\hat{\mathcal{H}},\hat{a}^{+}(\mathbf{k}',\nu')\right]=\sum_{\mathbf{k},\nu}\hbar\omega(\mathbf{k},\nu)\left\{\hat{a}^{+}(\mathbf{k},\nu)\hat{a}(\mathbf{k},\nu)\hat{a}^{+}(\mathbf{k}',\nu')\right.$$

$$\left.-\hat{a}^{+}(\mathbf{k}',\nu')\hat{a}^{+}(\mathbf{k},\nu)\hat{a}(\mathbf{k},\nu)\right\}$$

$$=\sum_{\mathbf{k},\nu}\hbar\omega(\mathbf{k},\nu)\left\{\hat{a}^{+}(\mathbf{k},\nu)\hat{a}(\mathbf{k},\nu)\hat{a}^{+}(\mathbf{k}',\nu')-\hat{a}^{+}(\mathbf{k},\nu)\hat{a}^{+}(\mathbf{k}',\nu')\hat{a}(\mathbf{k},\nu)\right\}$$

$$=\sum_{\mathbf{k},\nu}\hbar\omega(\mathbf{k},\nu)\hat{a}^{+}(\mathbf{k},\nu)\left[\hat{a}(\mathbf{k},\nu),\hat{a}^{+}(\mathbf{k}',\nu')\right]$$

$$=\hbar\omega(\mathbf{k}',\nu')\hat{a}^{+}(\mathbf{k}',\nu') \qquad \text{(G.11a)}$$

$$\left[\hat{\mathcal{H}},\hat{a}(\mathbf{k}',\nu')\right]=-\hbar\omega(\mathbf{k}',\nu')\hat{a}(\mathbf{k}',\nu') \qquad \text{(G.11b)}$$

In the second step in the development of equation (G.11a) we have used the commutation relations (G.10c) to allow us to reverse the order of the two operators.

FURTHER READING

Brüesch (1982) App. E

Appendix H

Published phonon dispersion curves for non-metallic crystals: update of previous compilation

Measurements of phonon dispersion curves for non-metallic crystals that have been published since 1984 and some that have been published during 1979–1983 are tabulated.

A compilation of the measured phonon dispersion curves, with references, for a number of metals is given by Willis and Pryor (1975, p 226). A similar compilation for insulators is given by Bilz and Kress (1979). The following tables update the compilation of phonon dispersion curves measured in non-metallic crystals. The references from 1984 have been extracted by searching through computer databases, but given that the searches are based on a choice of keywords it cannot be guaranteed that the list is exhaustive.

Sadly a large number of measured dispersion curves never get as far as publication! However, each neutron scattering institute publishes an annual report, which often contains such unpublished data. Alternative sources of dispersion curves are conference proceedings.

The references have been grouped under three headings: molecular crystals, silicates, and ionic crystals.

Molecular crystals

Material	Reference	Comments
Naphthalene, $C_{10}H_8$	Natkaniec et al. (1980)	Monoclinic; deuterated sample; all external modes and lowest-frequency internal modes measured along principal directions at 5 K
Anthracene, $C_{14}H_{10}$	Dorner et al. (1982)	Monoclinic; deuterated sample; all external modes and lowest-frequency

		internal modes measured along principal directions at 5 K
$C_2(CN)_4$	Chaplot et al. (1983)	Monoclinic; deuterated sample; all external modes and lowest-frequency internal modes measured along principal directions at 5 K
$C_6F_3Cl_3$	Dove et al. (1989)	Hexagonal; all external modes measured along three symmetry directions at 5 K
HCN	Mackenzie and Pawley (1979)	Tetragonal phase; deuterated sample; acoustic modes in **a***–**b*** plane at several temperatures above 160 K
Thiourea, $SC(NH_2)_2$	McKenzie (1975)	High-temperature orthorhombic phase; deuterated sample; acoustic modes and some optic modes (including soft transverse branch) along three symmetry directions at room temperature.
sym-triazine, $C_3N_3H_3$	Heilmann et al. (1979)	Rhombohedral phase; hydrogenated sample; acoustic modes along **b*** at several temperatures above 200 K
	Dove et al. (1983)	Rhombohedral phase; deuterated sample; acoustic modes along **b*** at several temperatures above 200 K; some optic modes
2,3-dimethyl-naphthalene, $C_{10}H_6(CH_3)_2$	Worlen et al. (1988)	Monoclinic phase; deuterated sample; acoustic and some optic modes along **a*** at 123 K
α-perylene, $C_{20}H_{12}$	Schleifer et al. (1989)	Monoclinic; deuterated sample; complete set of external modes and some internal modes along principal directions at 10 K

Silicates

Material	Reference	Comments
Quartz, SiO_2	Boysen et al. (1980); Dorner et al. (1980)	Trigonal phase; measurements along symmetry directions in basal plane at room temperature
	Berge et al. (1986); Dolino et al. (1992)	Hexagonal phase; measurements along symmetry directions in basal plane at several temperatures
Forsterite, Mg_2SiO_4	Rao et al. (1988)	Orthorhombic; measurements of lowest-frequency acoustic and optic modes along three symmetry directions at room temperature

Fayalite, Fe_2SiO_4	Ghose et al. (1991)	Orthorhombic
Andalusite, Al_2SiO_5	Winkler and Buehrer (1990)	Orthorhombic; acoustic modes and low-frequency optic modes along one direction at room temperature
Leucite, $KAlSi_2O_6$	Boysen (1990)	Cubic phase; acoustic modes and low-frequency optic modes measured along three symmetry directions at high temperature

Ionic crystals

Material	**Reference**	**Comments**
Calcite, $CaCO_3$	Cowley and Pant (1970)	Rhombohedral; almost a complete set of external modes along **c***
	Dove et al. (1992b)	Low-frequency modes along non-symmetry direction with soft mode for several temperatures
Corundum (Sapphire), Al_2O_3	Bialas and Stolz (1975)	Trigonal; acoustic and several optic branches along three-fold axis at room temperature
$NaNO_3$	Lefebvre et al. (1980)	Rhombohedral; complete set of external modes along **c***
$Cs_2NaBiCl_6$	Prokert and Aleksandrov (1984)	Cubic; acoustic modes and low-frequency optic modes along three symmetry directions at room temperature; soft optic branch for temperatures between 100 K and 300 K
$CoSi_2$	Weis et al. (1985)	CdI_2 structure; acoustic modes along three directions at room temperature
KSCN	Cookson et al. (1987)	Orthorhombic; acoustic and some optic modes along four directions at room temperature
$CaSO_4$	Schweiss et al. (1987)	Orthorhombic; acoustic and some optic modes along four directions at room temperature
$KTaO_3$	Perry et al. (1989)	Cubic; acoustic and several optic modes measured along three symmetry directions at room temperature; additional measurements repeated at temperatures over the range 4–1220 K
GaAs	Strauch and Dorner (1990)	Cubic; complete dispersion curves for three symmetry directions and zone boundary points at 10 K
Bi_2Te_3	Kullmann et al. (1990)	Rhombohedral; acoustic and some optic modes measured along three directions at 77 K

Li_2S	Buehrer et al. (1991)	Cubic; near-complete dispersion curves for three symmetry directions and zone boundary points at 10 K
La_2NiO_4	Pintschovius et al. (1989)	Tetragonal; complete dispersion curves for three symmetry directions and zone boundary points at room temperature
La_2CuO_4	Birgeneau et al. (1987)	Tetragonal; low-frequency modes along [110] at 423 K
	Pintschovius et al. (1991)	Near complete set of dispersion curves along [110] at 580 K
$La_{2-x}Sr_xCuO_4$	Böni et al. (1988)	Tetragonal; low-frequency modes along symmetry directions at several temperatures
	Pintschovius et al. (1991)	Complete set of dispersion curves along [110] at 295 K
Nd_2CuO_4	Pintschovius et al. (1991)	Tetragonal; complete set of dispersion curves along [100], [110] and [001] at room temperature
$YBa_2Cu_3O_6$	Pintschovius et al. (1991)	Near complete set of dispersion curves along [100]
$YBa_2Cu_3O_{6.85}$	Reichardt et al. (1989)	Acoustic modes and some optic modes along three directions.
$YBa_2Cu_3O_7$	Pintschovius et al. (1991)	Half-complete set of dispersion curves along [100]
$RbAlF_4$	Bulou et al. (1989)	Tetragonal; low-frequency modes at 673 K
U*X* (*X* = C, N, As, Sb, S, Se, Te)	Jackman et al. (1986)	Cubic NaCl structure; dispersion curves along all symmetry directions at room temperature. Some of the results were obtained by other workers as cited in this reference
$PdTe_2$	Finlayson et al. (1986)	CdI_2-type structure; complete dispersion curves along three symmetry directions at room temperature
$SnSe_2$	Harbec et al. (1983)	CdI_2-type structure; acoustic and some optic dispersion curves along three symmetry directions at room temperature
$MgSi_2$	Hutchings et al. (1988)	Cubic antifluoride; acoustic modes and some optic modes along three symmetry directions at room temperature

References

Adkins, C. J. (1975). *Equilibrium Thermodynamics* (second edition). London: McGraw-Hill.*

Aghdaee, S. R. & Rae, A. I. M. (1983). The phase transition in sodium azide. *J. Chem. Phys.* **79**, 4558–4563.

Alder, B. J. & Wainwright, T. W. (1957). Phase transition for a hard sphere system. *J. Chem. Phys.* **27**, 1208–1209.

Alder, B. J. & Wainwright, T. W. (1959). Studies in molecular dynamics. I. General method. *J. Chem. Phys.* **31**, 459–466.

Allen, M. P. & Tildesley, D. J. (1987). *Computer Simulation of Liquids*. Oxford: Clarendon.*

Ashcroft, N. W. & Mermin, N. D. (1976) *Solid State Physics*. New York: Holt, Rinehart & Winston.*

Axe, J. D. (1971). Neutron studies of displacive phase transitions. *Trans. Am. Cryst. Ass.* **7**, 89–103.

Bacon, G. E., ed. (1986). *Fifty Years of Neutron Diffraction: The Advent of Neutron Scattering*. Bristol: Hilger.*

Baldereschi, A. (1973). Mean value point in the Brillouin zone. *Phys. Rev.* **B7**, 5212–5215.

Barker Jr., A. S. & Tinkham, M. (1962). Far-infrared ferroelectric vibration mode in $SrTiO_3$. *Phys. Rev.* **125**, 1527–1530.

Barnes, R. B. (1932). Die Ultraroten Eigenfrequenzen der Alkalihalogenidkristalle.[1] *Zeit. Phys.* **75**, 723–734.

Barron, T. H. K., Collins, J. G. & White, G. K. (1980). Thermal expansion of solids at low temperatures. *Adv. Phys.* **29**, 609–730.

Bée, M. (1988). *Quasielastic Neutron Scattering: Principles and Applications in Solid State Chemistry, Biology and Materials Science*. Bristol: Hilger.*

Beeman, D. (1976). Some multistep methods for use in molecular dynamics calculations. *J. Comput. Phys.* **20**, 130–139.

Berge, B., Bachheimer, J. P., Dolino, J., Vallade, M. & Zeyen, C. (1986). Inelastic neutron scattering study of quartz near the incommensurate phase transition. *Ferroelectrics* **66**, 73–84.

Bethke, J. , Dolino, G., Eckold, G., Berge, B., Vallade, M., Zeyen, C., Hahn, T., Arnold, H. & Moussa, F. (1987). Phonon dispersion and mode coupling in high-

* An asterisk at the end of a reference indicates material for further reading.

[1] Translation: The infrared eigenfrequencies of alkalihalide crystals.

quartz near the incommensurate phase transition. *Europhys. Lett.* **3**, 207–212.
Bialas, H. & Stolz, H. J. (1975). Lattice dynamics of sapphire (corundum). *Zeit. Phys.* **B21**, 319–324.
Bilz, H. & Kress, W. (1979). *Phonon Dispersion Relations in Insulators*. Berlin: Springer-Verlag.*
Birgeneau, R. J., Chen, C. Y., Gabbe, D. R., Jenssen, H. P., Kastner, M. A., Peters, C. J., Picone, P. J., Thio, T., Thurston, T. R., Tuller, H. L., Axe, J. D., Böni, P. & Shirane, G. (1987). Soft-phonon behaviour and transport in single-crystal La_2CuO_4. *Phys. Rev.* **59**, 1329–1332.
Bismayer, U. (1988). New developments in Raman spectroscopy on structural phase transitions. In Salje, E. K. H. (ed.), *Thermodynamic and Physical Properties of Minerals*. Dordrecht: Riedel. pp 143–183.
Bismayer, U., Salje, E., Jansen, M. & Dreher, S. (1986). Raman scattering near the structural phase transition of As_2O_5: order parameter treatment. *J. Phys. C: Sol. State Phys.* **19**, 4537–4545.
Blinc, R. & Levanyuk, A. P., eds. (1986). *Incommensurate Phases in Dielectrics volume 2: Materials*. Amsterdam: North-Holland.*
Blinc, R. & Zeks, B. (1974). *Soft Modes in Ferroelectrics and Antiferroelectrics*. Amsterdam: North-Holland.*
Bohlen, S. R., Montana, A. & Kerrick, D. M. (1991). Precise determinations of the equilibria kyanite – sillimanite and kyanite – andalusite and a revised triple point for Al_2SiO_5 polymorphs. *Am. Min.* **76**, 677–680.
Böni, P., Axe, J. D., Shirane, G., Birgeneau, R. J., Gubbe, D. R., Jenssen, H. P., Kastner, M. A., Peters, C. J., Picone, P. J. & Thurston, T. R. (1988). Lattice instability and soft phonons in single-crystal $La_{2-x}Sr_xCuO_4$. *Phys. Rev.* **38**, 185–194.
Born, M. & Huang, K. (1954). *Dynamical Theory of Crystal Lattices*. Oxford: Oxford University Press.*
Born, M. & Mayer, J. E. (1932). Zur Gittertheorie der Ionenkristalle.[2] *Zeit. Phys.* **75**, 1–18.
Born, M. & Oppenheimer, J. (1927). Zur Quantentheorie der Molekeln.[3] *Ann. Phys.* **84**, 457–484.
Born, M. & von Kármán, Th. (1912). Über Schwingungen in Raumgittern.[4] *Phys. Zeit.* **13**, 297–309.
Born, M. & von Kármán, Th. (1913). Über die Verteilung der Eigenschwingungen von Punktgittern.[5] *Phys. Zeit.* **14**, 65–71.
Bounds, D. G., Klein, M. L. & Patey, G. N. (1980). Molecular dynamics simulation of the plastic phase of solid methane. *J. Chem. Phys.* **72**, 5348–5356.
Boysen, H. (1990). Neutron scattering and phase transitions in leucite. In Salje (1990). pp 334–349.
Boysen, H., Dorner, B., Frey, F. & Grimm, H. (1980). Dynamic structure determination for two interacting modes at the M point in α- and β-quartz by inelastic neutron scattering. *J. Phys. C: Sol. State Phys.* **13**, 6127–6146.
Bradley, C. J. & Cracknell, A. P. (1972). *The Mathematical Theory of Symmetry in Solids*. Oxford: Clarendon Press.*
Bragg, W. L. (1913). The structure of some crystals as indicated by their diffraction of X-rays. *Proc. Roy. Soc. London* **A89**, 248–277.

[2] Translation: On the lattice theory of ionic crystals.
[3] Translation: On the quantum theory of molecules.
[4] Translation: On the vibrations in three-dimensional lattices.
[5] Translation: On the distribution of vibrational modes in point lattices.

Brillouin, L. (1914). Diffusion de la lumière par un corps transparent homogène.[6] *Comptes Rendus* **158**, 1331–1334.

Brillouin, L. (1922). Diffusion de la lumière et des rayons X par un corps transparent homogène. Influence de l'agitation thermique.[7] *Ann. de Phys. (Paris)* **17**, 88–122.

Brillouin, L. (1930). Les électrons libre dans les métaux et le role des réflexions de Bragg.[8] *J. Phys. Radium* **1**, 377–400.

Brockhouse, B. N. (1961). Methods for neutron spectrometry. In *Inelastic Scattering of Neutrons in Solids and Liquids*. Vienna: IAEA. pp 113–151.

Brockhouse, B. N., Arase, T., Caglioti, G., Rao, K. R. & Woods, A. D. B. (1962). Crystal dynamics of lead. I. Dispersion curves at 100 K. *Phys. Rev.* **128**, 1099–1111.

Brockhouse, B. N. & Stewart, A. T. (1955). Scattering of neutrons by phonons in an aluminum single crystal. *Phys. Rev.* **100**, 756–757.

Bruce, A. D. & Cowley, R. A. (1973). Lattice dynamics of strontium titanate: anharmonic interactions and structural phase transitions. *J. Phys. C: Sol. State Phys.* **6**, 2422–2440.

Bruce, A. D. & Cowley, R. A. (1981). *Structural Phase Transitions*. London: Taylor & Francis Ltd.*

Brüesch, P. (1982). *Phonons: Theory and Experiments I. Lattice Dynamics and Models of Interatomic Forces*. Berlin: Springer-Verlag.*

Brüesch, P. (1986). *Phonons: Theory and Experiments II. Experiments and Interpretation of Experimental Results*. Berlin: Springer-Verlag.*

Brüesch, P. (1987). *Phonons: Theory and Experiments III. Phenomena related to Phonons*. Berlin: Springer-Verlag.*

Buehrer, W., Altorfer, F., Mesot, J., Bill, H., Carron, P. & Smith, H. G. (1991). Lattice dynamics and the diffuse phase transition of lithium sulfide investigated by coherent neutron scattering. *J. Phys.: Cond. Matter* **43**, 6202–6205.

Bulou, A., Rousseau, M., Nouet, J. & Hennion, B. (1989). Lattice dynamics and structural phase transitions in $RbAlF_4$: group theory, inelastic neutron scattering results and the calculation of the phonon spectrum. *J. Phys.: Cond. Matter* **1**, 4553–4583.

Burkel, E., Dorner, B., Illini, T. & Peisl, J. (1989). First observation of phonon dispersion curves with inelastic X-ray scattering. *Rev. Sci. Instr.* **60**, 1671–1673.

Burkel, E., Dorner, B., Illini, Th. & Peisl, J. (1991). High-energy resolution in X-ray scattering with the spectrometer INELAX. I. The principles and test instrument. *J. Appl. Cryst.* **24**, 1042–1050.

Burnham, C. W. (1990). The ionic model: perceptions and realities in mineralogy. *Am. Min.* **75**, 443–463.

Burns, G. & Scott, B. A. (1970). Raman studies of underdamped soft modes in $PbTiO_3$. *Phys. Rev. Lett.* **25**, 167–170.

Califano, S., Schettino, V. & Neto, N. (1981). *Lattice Dynamics of Molecular Crystals*. Berlin: Springer-Verlag.*

Castellano, E. E. & Main, P. (1985). On the classical interpretation of thermal probability ellipsoids and the Debye–Waller factor. *Acta Cryst.* **A41**, 156–157.

Chaplot, S. L., Mierzejewski, A., Pawley, G. S., Lefebvre, J. & Luty, T. (1983). Phonon dispersion of the external and low-frequency internal vibrations in monoclinic tetracyanoethylene at 5 K. *J. Phys. C: Sol. State Phys.* **16**, 625–644.

[6] Translation: Scattering of light by a homogenous transparent body.
[7] Translation: Scattering of light and X-rays by a homogenous transparent body. Effect of thermal excitations.
[8] Translation: Free electrons in metals and the role of Bragg reflections.

Chaplot, S. L., Pawley, G. S., Bokhenkov, E. L., Sheka, E. F., Dorner, B., Kalus, J., Jindal, V. K. & Natkaniec, I. (1981). Eigenvectors of low frequency internal phonons in crystalline anthracene-d_{10}. *Chem. Phys.* **57**, 407–414.

Chaplot, S. L., Pawley, G. S., Dorner, B., Jindal, V. K., Kalus, J. & Natkaniec, I. (1982). Calculated low frequency phonon dispersion in anthracene-d_{10}. *Phys. Stat. Sol.* **110b**, 445–454.

Cheung, P. S. Y. (1977). On the calculation of specific heats, thermal pressure coefficients and compressibilities in molecular dynamics simulations. *Mol. Phys.* **2**, 519–526.

Cheung, P. S. Y. & Powles, J. G. (1975). The properties of liquid nitrogen IV. A computer simulation. *Mol. Phys.* **30**, 921–949.

Cheung, P. S. Y. & Powles, J. G. (1976). The properties of liquid nitrogen V. Computer simulation with quadrupole interaction. *Mol. Phys.* **32**, 1383–1405.

Chihara, H., Nakamura, N. & Tachiki, M. (1973). Phase transition associated with a soft mode of molecular libration in crystal. *J. Chem. Phys.* **59**, 5387–5391.

Ciccotti, G., Frenkel, D. & McDonald, I. R., eds. (1987). *Simulations of Liquids and Solids: Molecular Dynamics and Monte Carlo Methods in Statistical Mechanics.* Amsterdam: North-Holland.*

Cochran, W. (1959a). Crystal stability and the theory of ferroelectricity. *Phys. Rev. Lett.* **3**, 412–414.

Cochran, W. (1959b). Dielectric constants and lattice vibrations of cubic ionic crystals. *Zeit. Krist.* **112**, 465–471.

Cochran, W. (1959c). Theory of the lattice vibrations of germanium. *Phys. Rev. Lett.* **2**, 495–497.

Cochran, W. (1960). Crystal stability and the theory of ferroelectricity. *Adv. Phys.* **9**, 387–423.

Cochran, W. (1961). Crystal stability and the theory of ferroelectricity Part II. Piezoelectric crystals. *Adv. Phys.* **10**, 401–420.

Cochran, W. (1971). Lattice dynamics of ionic and covalent crystals. *CRC Crit. Rev. Sol. State Sci.* **2**, 1–44.

Cochran, W. (1973). *The Dynamics of Atoms in Crystals.* London: Arnold.*

Cochran, W. (1981). Soft modes, a personal perspective. *Ferroelectrics* **35**, 3–8.

Cochran, W. & Cowley, R. A. (1962). Dielectric constants and lattice vibrations. *J. Phys. Chem. Sol.* **23**, 447–450.

Cochran, W. & Cowley, R. A. (1967). *Encyclopedia of Physics* 25/2a. Berlin: Springer.*

Cochran, W. & Pawley, G. S. (1964). The theory of diffuse scattering of X-rays by a molecular crystal. *Proc. Roy. Soc. London* **A280**, 1–22.

Cohen, R. E. (1991). Bonding and elasticity of stishovite SiO_2 at high pressure: linearised augmented plane wave calculations. *Am. Min.* **76**, 733–742.

Comès, R. & Shirane, G. (1972). Neutron scattering analysis of the linear displacement correlations in $KTaO_3$. *Phys. Rev.* **B5**, 1886–1891.

Cookson, D. J., Finlayson, T. R. & Elcome, M. M. (1987). Phonon dispersion relations for potassium thiocyanate. *Sol. State Comm.* **64**, 357–359.

Cowley, E. R. & Pant, A. K. (1970). Lattice dynamics of calcite. *Phys. Rev.* **B8**, 4795–4800.

Cowley, R. A. (1962). Temperature dependence of a transverse optic mode in strontium titanate. *Phys. Rev. Lett.* **9**, 159–161.

Cowley, R. A. (1976). Acoustic phonon instabilities and structural phase transitions. *Phys. Rev.* **B13**, 4877–4885.

Cowley, R. A., Buyers, W. L. & Dolling, G. (1969). Relationship of normal modes of vibration of strontium titanate and its antiferroelectric phase transition at 110 K. *Sol. State Comm.* **7**, 181–184.

Cowley, R. A., Woods, A. D. B. & Dolling, G. (1966). Crystal dynamics of potassium. I. Pseudopotential analysis of phonon dispersion curves at 9 K. *Phys. Rev.* **150**, 487–494.

Cox, S. R., Hsu, L.-Y. & Williams, D. E. (1981). Nonbonded potential function models for crystalline oxyhydrocarbons. *Acta Cryst.* **A37**, 293–301.

Daunt, S. J., Shurvell, H. F. & Pazdernik, L. (1975). The solid state vibrational spectra of s-triazine and s-triazine-d_3 and the monoclinic to rhombohedral phase transition. *J. Raman Spectr.* **4**, 205–223.

Debye, P. (1912). Zur Theorie der Spezifischen Wärmen.[9] *Ann. Phys.* **39**, 789–839.

Debye, P. (1914). Interferenz von Röntgenstrahlen und Wärmebewegung.[10] *Ann. Phys.* **43**, 49–95.

Dick, B. G., Jr. & Overhauser, A. W. (1958). Theory of the dielectric constants of alkali halide crystals. *Phys. Rev.* **112**, 90–103.

Dolino, G. (1990). The α–inc–β transitions of quartz: A century of research on displacive phase transitions. *Phase Transitions* **21**, 59–72.

Dolino, G., Berge, B., Vallade, M. & Moussa, F. (1989). Inelastic neutron scattering studies of the origin of the incommensurate phase of quartz. *Physica* **B156**, 15–16.

Dolino, G., Berge, B., Vallade, M. & Moussa, F. (1992). Origin of the incommensurate phase of quartz: I. Inelastic neutron scattering study of the high temperature β phase of quartz. *J. de Phys. I* **2**, 1461–1480.

Dolling, G. & Powell, B. M. (1970). Intermolecular dynamics of hexamethylenetetramine. *Proc. Roy. Soc. London* **A319**, 209–235.

Dolling, G., Powell, B. M. & Sears, V. F. (1979). Neutron diffraction study of the plastic phases of polycrystalline SF_6 and CBr_4. *Mol. Phys.* **37**, 1859–1883.

Dorner, B., ed. (1982). *Coherent Inelastic Neutron Scattering in Lattice Dynamics*. Berlin: Springer-Verlag.*

Dorner, B., Bokhenkov, E. L., Chaplot, S. E., Kalus, J., Natkaniec, I., Pawley, G. S., Schmelzer, U. & Sheka, E. F. (1982). The 12 external and the 4 lowest internal phonon dispersion branches in d_{10}-anthracene at 12 K. *J. Phys. C: Sol. State Phys.* **15**, 2353–2365.

Dorner, B., Burkel, E., Illini, T. & Peisl, J. (1987). First measurement of a phonon dispersion curve by inelastic X-ray scattering. *Zeit. Phys. B: Cond. Matter* **69**, 179–183.

Dorner, B., Grimm, H. & Rzany, H. (1980). Phonon dispersion branches in α-quartz. *J. Phys. C: Sol. State Phys.* **13**, 6607–6612.

Dove, M. T. (1988). Molecular dynamics simulations in the solid state sciences. In Salje, E. K. H. (ed.), *Thermodynamic and Physical Properties of Minerals*. Dordrecht: Riedel. pp 501–590.

Dove, M. T. (1989). On the computer modeling of diopside: toward a transferable potential model for silicate minerals. *Am. Min.* **74**, 774–779.

Dove, M. T., Giddy, A. P. & Heine, V. (1992a). On the application of mean-field and Landau theory to displacive phase transitions. *Ferroelectrics* **136**, 33–49.

Dove, M. T., Hagen, M. E., Harris, M. J., Powell, B. M., Steigenberger, U. & Winkler, B. (1992b) Anomalous inelastic neutron scattering from calcite. *J. Phys.: Cond. Matter* **4**, 2761–2774.

Dove, M. T., Heilmann, I. U., Kjems, J. K., Kurittu, J. & Pawley, G. S. (1983). Neutron scattering of phonons in per-deuterated s-triazine. *Phys. Stat. Sol.* **120b**, 173–181.

[9] Translation: On the theory of specific heat.

[10] Translation: Interference of X-rays and thermal motion.

Dove, M. T. & Lynden-Bell, R. M. (1986). A model of the paraelectric phase of thiourea. *Phil. Mag.* **B54**, 443–463.

Dove, M. T. & Pawley, G. S. (1984). A molecular dynamics simulation study of the orientationally disordered phase of sulphur hexafluoride. *J. Phys. C: Sol. State Phys.* **17**, 6581–6599.

Dove, M. T., Pawley, G. S., Dolling, G. & Powell, B. M. (1986). Collective excitations in an orientationally frustrated solid: Neutron scattering and computer simulation studies of SF_6. *Mol. Phys.* **57**, 865–880.

Dove, M. T. & Powell, B. M. (1989). Neutron diffraction study of the tricritical orientational order/disorder phase transition in calcite at 1260 K. *Phys. Chem. Min.* **16**, 503–507.

Dove, M. T., Powell, B. M., Pawley, G. S., Chaplot, S. L. & Mierzejewski, A. (1989). Inelastic neutron scattering determination of phonon dispersion curves in the molecular crystal sym-$C_6F_3Cl_3$. *J. Chem. Phys.* **90**, 1918–1923.

Dove, M. T., Winkler, B., Leslie, M., Harris, M. J. & Salje, E. K. H. (1992c). A new interatomic potential model for calcite: applications to lattice dynamics studies, phase transitions, and isotope fractionation. *Am. Min.* **17**, 244–250.

Dulong, P. L. & Petit, A. T. (1819). *Ann. Chem.* **10**, 395

Eckold, G., Stein-Arsic, M. & Weber, H. J. (1987). UNISOFT – a program package for lattice dynamical calculations. *J. Appl. Cryst.* **20**, 134–139.

Einstein, A. (1907). Die Plancksche Theorie der Strahlung und die Theorie der Spezifischen Wärme.[11] *Ann. Phys.* **22**, 180–190.

Elcombe, M. M. & Pryor, A. W. (1970). The lattice dynamics of calcium fluoride. *J. Phys. C: Sol. State Phys.* **3**, 492–499.

Endoh, Y., Shirane, G. & Skalyo, J. Jr. (1975). Lattice dynamics of solid neon at 6.5 and 23.7 K. *Phys. Rev.* **B11**, 1681–1688.

Ewald, P. P. (1921). Die Berechnung optischer und elektrostatische Gitterpotentiale.[12] *Ann. Phys.* **64**, 253–287.

Finlayson, T. R., Reichardt, W. & Smith, H. G. (1986). Lattice dynamics of layered structure compounds: $PdTe_2$. *Phys. Rev.* **B33**, 2473–2480.

Fisher, M. E. (1983). Scaling, universality and renormalisation group theory. In Hahne, F. J. W. (ed.) *Critical Phenomena*. Berlin: Springer-Verlag. pp 1–139.

Fleury, P. A., Scott, J. F. & Worlock, J. M. (1968). Soft phonon modes and the 110 K phase transition in $SrTiO_3$. *Phys. Rev. Lett.* **21**, 16–19.

Fujii, Y., Lurie, N. A., Pynn, R. & Shirane, G. (1974). Inelastic neutron scattering from solid ^{36}Ar. *Phys. Rev.* **B10**, 3647–3659.

Ghose, S., Hastings, J. M., Choudhury, N., Chaplot, S. L. & Rao, K. R. (1991). Phonon dispersion relation in fayalite, Fe_2SiO_4. *Physica* **B174**, 83–86.

Giddy, A. P., Dove, M. T. & Heine, V. (1989). What do Landau free energies really look like for structural phase transitions? *J. Phys.: Cond. Matter* **1**, 8327–8335.

Giddy, A. P., Dove, M. T. & Heine, V. (1990). The non-analytic nature of Landau free energies. *Ferroelectrics* **104**, 331–335.

Gross, E. (1932a) Change of wavelength of light due to elastic heat waves at scattering in liquids. *Nature* **126**, 201–202.

Gross, E. (1932b) The splitting of spectral lines at scattering of light by liquids. *Nature* **126**, 400.

Gross, E. (1932c) Splitting of the frequency of light scattered by liquids and optical anisotropy of molecules. *Nature* **126**, 603–604.

[11] Translation: The Planck theory of radiation and the theory of specific heat.
[12] Translation: The calculation of optical and electrostatic lattice potentials.

Gross, E. (1932d) Modification of light quanta by elastic heat oscillations in scattering media. *Nature* **129**, 722–723.

Güttler, B., Salje, E. & Putnis, A. (1989). Structural states in Mg-cordierite III: infrared spectroscopy and the nature of the hexagonal–modulated transition. *Phys. Chem. Min.* **16**, 365–373.

Hansen, J. P. & Klein, M. L. (1976). Dynamical structure factor $S(\mathbf{Q}, \omega)$ of solid potassium. *Sol. State Comm.* **20**, 771–773.

Harada, J., Axe, J. D. & Shirane, G. (1971). Neutron scattering study of soft modes in cubic $BaTiO_3$. *Phys. Rev.* **B4**, 155–162.

Harbec, J. Y., Powell, B. M. & Jandl, S. (1983). Lattice dynamics of $SnSe_2$. *Phys. Rev.* **B28**, 7009–7013.

Harris, M. J. (1992) Ferroelectric ordering in $NaNO_3$. *Sol. State Comm.* **84**, 557–561.

Harris, M. J., Salje, E. K. H. & Güttler, B. K. (1990). An infrared spectroscopic study of the internal modes of sodium nitrate: implications for the structural phase transition. *J. Phys.: Cond. Matt.* **2**, 5517–5527.

Heilmann, I. U., Ellenson, W. D. & Eckerd, J. (1979). Softening of transverse elastic modes at the structural phase transition of s-triazine, $C_3N_3H_3$. *J. Phys. C: Sol. State Phys.* **12**, L185–L189.

Hemingway, B. S., Robie, R. A., Evans, H. T. Jr. & Kerrick, D. M. (1991). Heat capacities and entropies of sillimanite, fibrolite, andalusite, kyanite, and quartz and the Al_2SiO_5 phase diagram. *Am. Min.* **76**, 1597–1613.

Hemley, R. J., Jackson, M. D. & Gordon, R. G. (1987). Theoretical study of the structures, lattice dynamics, and equations of state of perovskite-type $MgSiO_3$ & $CaSiO_3$. *Phys. Chem. Min.* **14**, 2–12.

Hofmann, W., Kalus, J., Lauterback, B., Schmelzer, U. & Selback, J. (1992). Measurement of acoustic phonons in beryllium by inelastic X-ray scattering. *Zeit. für Phys.* **B88**, 169–172.

Hsu, L.-Y. & Williams, D. E. (1980). Intermolecular potential-function models for crystalline perchlorohydrocarbons. *Acta Cryst.* **A36**, 277–281.

Hutchings, M. T., Farley, T. W. D., Kackett, M. A., Hayes, W., Hull, S. & Steigenberger, U. (1988). Neutron scattering investigation of lattice dynamics and thermally induced disorder in the antifluorite Mg_2Si. *Sol. State Ionics* **28**, 1208–1212.

Iizumi, M., Axe, J. D., Shirane, G. & Shimaoka, K. (1977). Structural phase transformation in K_2SeO_4. *Phys. Rev.* **B15**, 4392–4411.

Inkson, J. C. (1984). *Many-body Theory of Solids: An Introduction*. New York: Plenum.*

Iqbal, Z. & Owens, F. J., eds. (1984). *Vibrational Spectroscopy of Phase Transitions*. Orlando: Academic Press.*

Jackman, J. A., Holden, T. M., Buyers, W. J. L., DuPlessis, P. D., Vogt, O. & Genossar, J. (1986). Systematic study of the lattice dynamics of the uranium rocksalt structure compounds. *Phys. Rev.* **B33**, 7144–7153.

Jackson, H. E. & Walker, C. T. (1971). Thermal conductivity, second sound, and phonon–phonon interactions in NaF. *Phys. Rev.* **B3**, 1428–1439.

Jackson, R. A. & Catlow, C. R. A. (1988). Computer simulation studies of zeolite structures. *Mol. Sim.* **1**, 207–224.

Jacucci, G., McDonald, I. R. & Rahman, A. (1976). Effects of polarisation on equilibrium and dynamic properties of ionic systems. *Phys. Rev.* **A13**, 1581–1592.

Kellerman, E. W. (1940). Theory of the vibrations of the sodium chloride lattice. *Phil. Trans. Roy. Soc.* **238**, 513–548.

Kieffer, S.-W. (1979a). Thermodynamics and lattice vibrations of minerals, I: Mineral heat capacities and their relationship to simple lattice vibrational modes. *Rev. Geophys. Space Phys.* **17**, 1–19.

Kieffer, S.-W. (1979b). Thermodynamics and lattice vibrations of minerals, II: Vibrational characteristics of silicates. *Rev. Geophys. Space Phys.* **17**, 20–34.

Kieffer, S.-W. (1979c). Thermodynamics and lattice vibrations of minerals, III: lattice dynamics and an approximation for minerals with applications to simple substances and framework silicates. *Rev. Geophys. Space Phys.* **17**, 35–59.

Kieffer, S.-W. (1980). Thermodynamics and lattice vibrations of minerals, IV: application to chain and sheet silicates and orthosilicates. *Rev. Geophys. Space Phys.* **18**, 862–886.

Kittel, C. (1976). *Introduction to Solid State Physics* (fifth edition). New York: John Wiley.*

Klein, M. L. (1978). Computer simulation of collective modes in solids. In *Computer Modelling of Matter*, ed. P. Lykos. Washington: ACS. pp 94–110.

Klein, M. L. & Weiss, J. J. (1977). The dynamical structure factor $S(\mathbf{Q}, \omega)$ of solid β-N_2. *J. Chem. Phys.* **67**, 217–224.

Kramer, G. J., Farragher, N. P., van Beest, B. W. H. & van Santen, R. A. (1991). Interatomic force fields for silicates, aluminophosphates, and zeolites: derivation based on *ab initio* calculations. *Phys. Rev.* **B43**, 5068–5079.

Kuhs, W. F. (1992). Generalised atomic displacements in crystallographic structure analysis. *Acta Cryst.* **A48**, 80–98.

Kullmann, W., Eichhorn, G., Rauh, H., Geick, R., Eckold, G. & Steigenberger, U. (1990). Lattice dynamics and phonon dispersion in the narrow gap semiconductor Bi_2Te_3 with sandwich structure. *Phys. Stat. Sol.* **126b**, 125–140.

Landau, L. D. (1937). Zur Theorie der Phasenumwandlunger.[13] *Phys. Z. Sowjetunion.* **11**, 26–47.

Landau, L. D. & Lifshitz, E. M. (1980). *Statistical Physics*. Oxford: Pergamon Press.*

Landsberg, G. & Mandelstam, L. (1928). Eine neue Erscheinung bei der Lichtzerstreuung in Kristallen.[14] *Naturwiss.* **16**, 557–558.

Landsberg, G. & Mandelstam, L. (1929). Lichtzerstreuung in Kristallen bei hoher Temperatur.[15] *Zeit. Phys.* **58**, 250.

Lasaga, A. C. & Gibbs, G. V. (1987). Applications of quantum mechanical potential surfaces to mineral physics calculations. *Phys. Chem. Min.* **14**, 107–117.

Lasaga, A. C. & Gibbs, G. V. (1988). Quantum mechanical potential surfaces and calculations on minerals and molecular clusters. *Phys. Chem. Min.* **16**, 29–41.

Lazarev, A. N. & Mirgorodsky, A. P. (1991). Molecular force constants in dynamical model of α-quartz. *Phys. Chem. Min.* **18**, 231–243.

Lebowitz, J. L., Percus, J. K. & Verlet, L. (1967). Ensemble dependence of fluctuations with application to machine computations. *Phys. Rev.* **153**, 250–254.

Lefebvre, J., Currat, R., Fouret, R. & More, M. (1980). Etude par diffusion neutronique des vibrations de reseau dans le nitrate de sodium.[16] *J. Phys. C: Sol. State Phys.* **13**, 4449–4461.

Leinenweber, K. & Navrotsky, A. (1988). A transferable interatomic potential for crystalline phases in the system MgO–SiO_2. *Phys. Chem. Min.* **15**, 588–596.

Lines, M. E. & Glass, A. M. (1977). *Principles and Applications of Ferroelectrics and Related Materials*. Oxford: Clarendon Press.*

[13] Translation: On the theory of phase transitions.
[14] Translation: A new phenomenon in the light scattering in crystals.
[15] Translation: Light scattering in crystals at high temperatures.
[16] Translation: Neutron scattering studies of lattice vibrations in sodium nitrate.

London, F. (1930). Über einige Eigenschaften und Anwendungen der Molekulerkräfte.[17] *Zeit. Phys. Chem.* **B11**, 222–251.

Lovesey, S. W. (1984). *Theory of Neutron Scattering from Condensed Matter, Volume 1*. Oxford: Clarendon Press.*

Lurie, N. A., Shirane, G. & Skalyo, J. Jr. (1974). Phonon dispersion relations in xenon at 10 K. *Phys. Rev.* **B9**, 5300–5306.

Luty, T. & Pawley, G. S. (1974). A shell model for molecular crystals. *Phys. Stat. Sol.* **66b**, 309–319.

Luty, T. & Pawley, G. S. (1975). A shell model for molecular crystals: Orthorhombic sulphur. *Phys. Stat. Sol.* **69b**, 551–555.

Lyddane, R. H., Sachs, R. G. & Teller, E. (1941). On the polar vibrations of alkali halides. *Phys. Rev.* **59**, 673–676.

Mackenzie, G. A. & Pawley, G. S. (1979). Neutron scattering study of DCN. *J. Phys. C: Sol. State Phys.* **12**, 2717–2735.

Madelung, E. (1910). Molekulare Eigenschwingungen.[18] *Phys. Zeit.* **11**, 898–905.

Madelung, E. (1918). Das elektrische Feld in Systemen von regelmässig angeordneten Punktladungen.[19] *Phys. Zeit.* **19**, 524–532.

Mandl, F. (1971). *Statistical Physics*. Chichester: John Wiley.*

Maradudin, A. A. & Vosko, S. H. (1968). Symmetry properties of the normal vibrations of a crystal. *Rev. Mod. Phys.* **40**, 1–37.

Marshall, W. & Lovesey, S. W. (1971). *Theory of Thermal Neutron Scattering*. Oxford: Clarendon Press.*

Matsui, M. (1988). Molecular dynamics study of $MgSiO_3$ perovskite. *Phys. Chem. Min.* **16**, 234–238.

Matsui, M. (1989). Molecular dynamics study of the structural and thermodynamic properties of MgO crystal with quantum correction. *J. Chem. Phys.* **91**, 489–494.

Matsui, M. & Price, G. D. (1992). Computer simulations of the $MgSiO_3$ polymorphs. *Phys. Chem. Min.* **18**, 365–372.

McKenzie, D. R. (1975). Neutron and Raman study of the lattice dynamics of deuterated thiourea. *J. Phys. C: Sol. State Phys.* **8**, 2003–2010.

McMillan, P. F. & Hess, A. C. (1990). *Ab initio* valence force field calculations for quartz. *Phys. Chem. Min.* **17**, 97–107.

McMillan, P. F. & Hofmeister, A. M. (1988). Infrared and Raman spectroscopy. *Rev. Min.* **18**, 99–159.

Müller, K. A. & Berlinger, W. (1971). Static critical exponents at structural phase transitions. *Phys. Rev. Lett.* **26**, 13–16.

Nada, R., Catlow, C. R. A., Dovesi, R. & Pisani, C. (1990). An *ab initio* Hartree–Fock study of α-quartz and stishovite. *Phys. Chem. Min.* **17**, 353–362.

Natkaniec, I., Bokhenkov, E. L., Dorner, B., Kalus, J., Mackenzie, G. A., Pawley, G. S., Schmelzer, U. & Sheka, E. F. (1980). Phonon dispersion in d_8-naphthalene crystal at 6 K. *J. Phys. C: Sol. State Phys.* **13**, 4265–4283.

Normand, B. N. H., Giddy, A. P., Dove, M. T. & Heine, V. (1990). Bifurcation behaviour in structural phase transitions with double well potentials. *J. Phys.: Cond. Matter* **2**, 3737–3746.

Nosé, S. (1984a). A molecular dynamics method for simulations in the canonical ensemble. *Mol. Phys.* **52**, 255–268.

Nosé, S. (1984b). A unified formulation of the constant temperature molecular dynamics methods. *J. Chem. Phys.* **81**, 511–519.

[17] Translation: On some properties and applications of molecular forces.

[18] Translation: Molecular vibrations.

[19] Translation: The electric field in systems of periodically ordered point charges.

Nosé, S. & Klein, M. L. (1983). Constant pressure molecular dynamics for molecular systems. *Mol. Phys.* **50**, 1055–1076.
Nye, J. F. (1964). *Physical Properties of Crystals*. Oxford: Clarendon Press.*
Parker, S. C. & Price, G. D. (1989). Computer modelling of phase transitions in minerals. *Adv. Sol. State Chem.* **1**, 295–327.
Parlinski, K. (1988). Molecular dynamics simulation of incommensurate phases. *Comp. Phys. Rep.* **8**, 153–219.
Parrinello, M. & Rahman, A. (1980). Crystal structure and pair potentials: a molecular dynamics study. *Phys. Rev. Lett.* **45**, 1196–1199.
Parrinello, M. & Rahman, A. (1981). Polymorphic transitions in single crystals: a new molecular dynamics method. *J. Appl. Phys.* **52**, 7182–7190.
Patel, A., Price, G. D. & Mendelssohn, M. J. (1991). A computer simulation approach to modeling the structure, thermodynamics and oxygen isotope equilibria of silicates. *Phys. Chem. Min.* **17**, 690–699.
Pawley, G. S. (1967). A model for the lattice dynamics of naphthalene and anthracene. *Phys. Stat. Sol.* **20**, 347–360.
Pawley, G. S. (1972). Analytic formulation of molecular lattice dynamics based on pair potentials. *Phys. Stat. Sol.* **49b**, 475–488.
Pawley, G. S. & Leech, J. W. (1977). A lattice dynamical shell model for crystals with molecules having fixed multipole moments. *J. Phys. C: Sol. State Phys.* **10**, 2527–2546.
Pawley, G. S., MacKenzie, G. A., Bokhenkov, E. L., Sheka, E. F., Dorner, B., Kalus, J., Schmelzer, U. & Natkaniec, I. (1980). Determination of phonon eigenvectors in naphthalene by fitting neutron scattering intensities. *Mol. Phys.* **39**, 251–260.
Perry, C. H., Currat, R., Buhay, H., Migoni, R. M., Stirling, W. G. & Axe, J. D. (1989). Phonon dispersion and lattice dynamics of $KTaO_3$ from 4 K to 1220 K. *Phys. Rev.* **B39**, 8666–8676.
Petzelt, J. & Dvorak, V. (1976a). Changes of infrared and Raman spectra induced by structural phase transitions: I. General considerations. *J. Phys. C: Sol. State Phys.* **9**, 1571–1586.
Petzelt, J. & Dvorak, V. (1976b). Changes of infrared and Raman spectra induced by structural phase transitions: II. Examples. *J. Phys. C: Sol. State Phys.* **9**, 1587–1601.
Petzelt, J. & Dvorak, V. (1984). Infrared spectroscopy of structural phase transitions in crystals. In Iqbal and Owens (1984). pp 56–153.
Pintschovius, L., Bassat, J. M., Odier, P., Gervais, F., Chevrier, G., Reichardt, W. & Gompf, F. (1989). Lattice dynamics of La_2NiO_4. *Phys. Rev.* **B40**, 2229–2238.
Pintschovius, L., Pyka, N., Reichardt, W., Rumiantsev, A. Y., Mitrofanov, N. L., Ivanov, A. S., Collin, G. & Bourges, P. (1991). Lattice dynamic studies of HTSC materials. *Physica* **C185**, 156–161.
Placzek, G. & Van Hove, L. (1954). Crystal dynamics and inelastic scattering of neutrons. *Phys. Rev.* **93**, 1207–1214.
Poon, W. C.-K., Putnis, A. & Salje, E. (1990). Structural states of Mg-cordierite IV: Raman spectroscopy and local order parameter behaviour. *J. Phys.: Cond. Matt.* **2**, 6361–6372.
Post, J. E. & Burnham, C. W. (1986). Ionic modeling of mineral structures and energies in the electron-gas approximation: TiO_2 polymorphs, quartz, forsterite, diopside. *Am. Min.* **71**, 142–150.
Price, G. D., Parker, S. C. & Leslie, M. (1987a). The lattice dynamics of forsterite. *Min. Mag.* **51**, 157–170.
Price, G. D., Parker, S. C. & Leslie, M. (1987b). The lattice dynamics and thermodynamics of the Mg_2SiO_4 polymorphs. *Phys. Chem. Min.* **15**, 181–190.

Prokert, F. & Aleksandrov, K. S. (1984). Neutron scattering studies on phase transition and phonon dispersion in $Cs_2NaBiCl_6$. *Phys. Stat. Sol.* **124b**, 503–513.
Purton, J. & Catlow, C. R. A. (1990). Computer simulation of feldspar structures. *Am. Min.* **75**, 1268–1273.
Putnis, A., Salje, E., Redfern, S., Fyfe, C. & Strobel, H. (1987). Structural states of Mg-cordierite I: Order parameters from synchrotron, X-ray and NMR data. *Phys. Chem. Min.* **14**, 446–454.
Rae, A. I. M. (1981). *Quantum Mechanics*. London: McGraw-Hill.*
Rae, A. I. M. (1982). The structural phase change in s-triazine – the quasiharmonic approximation. *J. Phys. C: Sol. State Phys.* **15**, 1883–1896.
Rahman, A. (1964). Correlations in the motion of atoms in liquid argon. *Phys. Rev.* **136**, A405–A411.
Rahman, A. (1966). Liquid structure and self-diffusion. *J. Chem. Phys.* **45**, 2585–2592.
Rahman, A. & Stillinger, F. H. (1971). Molecular dynamics study of liquid water. *J. Chem. Phys.* **55**, 3336–3359.
Rahman, A. & Stillinger, F. H. (1974). Propagation of sound in water. A molecular dynamics study. *Phys. Rev.* **A10**, 368–378.
Raman, C. V. (1928). A new radiation. *Ind. J. Phys.* **2**, 387–398.
Raman, C. V. & Krishnan, K. S. (1928a). A new type of secondary radiation. *Nature* **121**, 501–502.
Raman, C. V. & Krishnan, K. S. (1928b). A new class of spectra due to secondary radiation. Part I. *Ind. J. Phys.* **2**, 399–419.
Raman, C. V. & Nedungadi, T. M. K. (1940). The α–β transformation of quartz. *Nature* **145**, 147.
Rao, K. R., Chaplot, S. L., Choudhury, N., Ghose, S., Hastings, J. M., Corliss, L. M. & Price, D. L. (1988). Lattice dynamics and inelastic neutron scattering from forsterite, Mg_2SiO_4: phonon dispersion relations, density of states, and specific heat. *Phys. Chem. Min.* **16**, 83–97.
Rao, C. N. R. & Rao, K. J. (1978). *Phase Transitions in Solids*. New York: McGraw–Hill.*
Raunio, G., Almqvist, L. & Stedman, R. (1969). Phonon dispersion relations in NaCl. *Phys. Rev.* **178**, 1496–1501.
Reichardt, W., Ewert, D., Gering, E., Gompf, F., Pintschovius, L., Renker, B., Collin, G., Dianoux, A. J. & Mutka, H. (1989). Lattice dynamics of 123 superconductors. *Physica* **B156**, 897–901.
Reynolds, P. A. (1973). Lattice dynamics of the pyrazine crystal structure by coherent inelastic neutron scattering. *J. Chem. Phys.* **59**, 2777–2786.
Righini, R., Califano, S. & Walmsley, S. M. (1980). Calculated phonon dispersion curves for fully deuterated naphthalene crystals at low temperature. *Chem. Phys.* **50**, 113–117.
Riste, T., Samuelsen, E. J., Otnes, K. & Feder, J. (1971). Crystal behaviour of $SrTiO_3$ near the 105 K phase transition. *Sol. State Comm.* **9**, 1455–1458.
Robie, R. A. & Hemingway, B. S. (1984). Entropies of kyanite, andalusite and sillimanite: additional constraints on the pressure and temperature of the Al_2SiO_5 triple point. *Am. Min.* **69**, 289–306.
Rossman, G. R. (1988). Vibrational spectroscopy of hydrous components. *Rev. Min.* **18**, 193–206.
Rubens, H. & Hollnagel, P. (1910). *Sitz. Berl. Akad.* **26**
Saksena, B. D. (1940). Analysis of the Raman and infrared spectra of α–quartz. *Proc. Ind. Acad. Sci.* **A12**, 93–139.
Salje, E. K. H. (1990). *Phase Transitions in Ferroelastic and Co-elastic Crystals*. Cambridge: Cambridge University Press.*

Salje, E. K. H. (1992). Hard mode spectroscopy: experimental studies of structural phase transitions. *Phase Transitions* **37**, 83–110.
Salje, E. K. H., Ridgewell, A., Güttler, B., Wruck, B., Dove, M. T. & Dolino, G. (1992). On the displacive character of the phase transition in quartz: a hard mode spectroscopy study. *J. Phys.: Cond. Matt.* **4**, 571–577.
Salje, E. & Werneke, C. (1982). The phase equilibrium between sillimanite and andalusite as determined from lattice vibrations. *Cont. Min. Pet.* **79**, 56–76.
Salje, E., Wruck, B. & Thomas, H. (1991). Order parameter saturation and a low-temperature extension of Landau theory. *Z. Phys: Cond. Matter* **82**, 399–404.
Sanders, M. J., Leslie, M. & Catlow, C. R. A. (1984). Interatomic potentials for SiO_2. *J. Chem. Soc., Chem. Comm.* 1271–1273.
Sangster, M. J. L. & Atwood, R. M. (1978). Ionic potentials for alkali halides: II Completely crystal independent specification of Born–Mayer potentials. *J. Phys. C: Sol. State Phys.* **11**, 1541–1555.
Sangster, M. J. L., Peckham, G. & Saunderson, D. H. (1970). Lattice dynamics of magnesium oxide. *J. Phys. C: Sol. State Phys.* **3**, 1026–1036.
Sangster, M. J. L., Schröder, U. & Attwood, R. M.. (1978). Ionic potentials for alkali halides: I Crystal independent shell parameters and fitted Born–Mayer potentials. *J. Phys. C: Sol. State Phys.* **11**, 1523–1541.
Schleifer, J., Kalus, J., Schmelzer, U. & Eckold, G. (1989). Phonon dispersion in an α-perylene-d_{12} crystal at 10 K. *Phys. Stat. Sol.* **154b**, 153–166.
Schmahl, W. W., Swainson, I. P., Dove, M. T. & Graeme-Barber, A. (1992). Landau free energy and order parameter behaviour of the α/β phase transition in cristobalite. *Zeit. Krist.* **201**, 124–145.
Schneider, T. & Stoll, E. (1973). Molecular dynamics investigation of structural phase transitions. *Phys. Rev. Lett.* **31**, 1254–1258.
Schneider, T. & Stoll, E. (1975). Observation of cluster waves and their lifetime. *Phys. Rev. Lett.* **35**, 296–299.
Schneider, T. & Stoll, E. (1976). Molecular dynamics study of structural phase transitions. I. One-component displacement models. *Phys. Rev.* **B13**, 1216–1237.
Schneider, T. & Stoll, E. (1978). Molecular dynamics study of a three-dimensional one-component model for distortive phase transitions. *Phys. Rev.* **B17**, 1302–1322.
Schweiss, B. P., Dyck, W. & Fuess, H. (1987). A theoretical lattice dynamics model and phonon dispersion measurements for $CaSO_4$ (anhydrite). *J. Phys. C: Sol. State Phys.* **20**, 651–670.
Scott, J. F. (1968). Evidence of coupling between one- and two-phonon excitations in quartz. *Phys. Rev. Lett.* **21**, 907–910.
Scott, J. F. (1974). Soft mode spectroscopy: experimental studies of structural phase transitions. *Rev. Mod. Phys.* **46**, 83–128.
Sears, V. F. (1986). Neutron scattering lengths and cross sections. In Skold and Price (1986). pp 521–550.
Sears, V. F. (1992). Neutron scattering lengths and cross sections. *Neutron News.* **3**, 26–37.
Shapiro, S. M., O'Shea, D. C. & Cummins, H. Z. (1967). Raman scattering study of the α–β phase transition in quartz. *Phys. Rev. Lett.* **19**, 361–364.
Shirane, G. (1974). Neutron scattering studies of structural phase transitions at Brookhaven. *Rev. Mod. Phys.* **46**, 437–449.
Shirane, G., Axe, J. D., Harada, J. & Remeika, J. P. (1970). Soft ferroelectric modes in lead titanate. *Phys. Rev.* **B2**, 155–159.
Shirane, G. & Yamada, Y. (1969). Lattice-dynamical study of the 110 K phase transition in $SrTiO_3$. *Phys. Rev.* **177**, 858–863.

Singer, K., Taylor, A. & Singer, J. V. L. (1977). Thermodynamic and structural properties of liquids modelled by '2-Lennard-Jones centres' pair potentials. *Mol. Phys.* **33**, 1757–1795.

Skalyo, J. Jr., Endoh, Y. & Shirane, G. (1974). Inelastic neutron scattering from solid krypton at 10 K. *Phys. Rev.* **B9**, 1797–1803.

Skold, K. & Price, D. L., eds. (1986). *Neutron Scattering*. Orlando: Academic Press.*

Squires, G. (1978). *Introduction to the Theory of Thermal Neutron Scattering*. Cambridge: Cambridge University Press.*

Steigenberger, U., Hagen, M., Caciuffo, R., Petrillo, C., Cilloco, F. & Sachetti, F. (1991). The development of the PRISMA spectrometer at ISIS. *Nucl. Instr. Methods* **B53**, 87–96.

Stirling, W. G. (1972). Neutron inelastic scattering study of the lattice dynamics of strontium titanate: harmonic models. *J. Phys. C: Sol. State Phys.* **5**, 2711–2730.

Stixrude, L. & Bukowinski, M. S. T. (1988). Simple covalent potential models of tetragonal SiO_2: applications to α-quartz and coesite at pressure. *Phys. Chem. Min.* **16**, 199–206.

Strauch, D. & Dorner, B. (1990). Phonon dispersion in GaAs. *J. Phys: Cond. Matter* **2**, 1457–1474.

Swainson, I. P. & Dove, M. T. (1992). Low frequency floppy modes in β-cristobalite. *Phys. Rev. Lett.* (in press).

Tautz, F. S., Heine, V., Dove, M. T. & Chen. X. (1991). Rigid unit modes in the molecular dynamics simulation of quartz and the incommensurate phase transition. *Phys. Chem. Min.* **18**, 326–336.

Terhune, R. W., Kushida, T. & Ford, G. W. (1985). Soft acoustic modes in trigonal crystals. *Phys. Rev.* **B32**, 8416–8419.

Toledano, J. C. & Toledano, P. (1988). *The Landau Theory of Phase Transitions*. Singapore: World Scientific.*

Traylor, J. G., Smith, H. G., Nicklow, R. M. & Wilkinson, M. K. (1971). Lattice dynamics of rutile. *Phys. Rev.* **B3**, 3457–3472.

Tsuneyuki, S., Aoki, H. & Tsukada, M. (1990). Molecular dynamics study of the α to β structural phase transition of quartz. *Phys. Rev. Lett.* **64**, 776–779.

Tsuneyuki, S., Tsukada, M., Aoki, H. & Matsui, Y. (1988). First principles interatomic potential of silica applied to molecular dynamics. *Phys. Rev. Lett.* **61**, 869–872.

Vallade, M., Berge, B. & Dolino, G. (1992). Origin of the incommensurate phase of quartz: II. Interpretation of inelastic neutron scattering data. *J. de Phys. I.* **2**, 1481–1495.

Van Hove, L. (1954). Correlations in space and time and Born approximation scattering in systems of interacting particles. *Phys. Rev.* **95**, 249–262.

Verlet, L. (1967). Computer "experiments" on classical fluids I: Thermodynamical properties of Lennard-Jones molecules. *Phys. Rev.* **159**, 98–103.

Walker, C. B. (1956). X-ray study of lattice vibrations in aluminum. *Phys. Rev.* **103**, 547–561.

Waller, I. (1923). Zur Frage der Einwirkung der Wärmebewegung auf die Interferenz von Röntgenstrahlen.[20] *Zeit. Phys.* **17**, 398–408.

Waller, I. (1928). Über eine verallgemeinerte Streuungsformel.[21] *Zeit. Phys.* **51**, 213–231.

Warren, J. L. (1968). Further consideration on the symmetry properties of the normal vibrations of a crystal. *Rev. Mod. Phys.* **40**, 38–76.

[20] Translation: On the question of the effect of thermal motion on X-ray interference.
[21] Translation: On a generalised scattering formula.

Warren, J. L. & Worton, T. G. (1974). Improved version of group-theoretical analysis of lattice dynamics. *Comp. Phys. Comm.* **8**, 71–84.
Weis, J. J. & Klein, M. L. (1975). The dynamical structure factor $S(\mathbf{Q}, \omega)$ of solid α-N_2. *J. Chem. Phys.* **63**, 2869–2873.
Weis, L., Rumyantsev, A. Y. & Ivanov, A. S. (1985). Acoustic phonon dispersion of $CoSi_2$. *Phys. Stat. Sol.* **128b**, K111–K115.
Williams, D. E. (1966). Nonbonded potential parameters derived from crystalline aromatic hydrocarbons. *J. Chem. Phys.* **45**, 3770–3778.
Williams, D. E. (1967). Nonbonded potential parameters derived from crystalline hydrocarbons. *J. Chem. Phys.* **47**, 4680–4684.
Williams, D. E. (1971). Accelerated convergence of crystal-lattice potential sums. *Acta Cryst.* **A27**, 452–455.
Williams, D. E. (1972). Molecular packing analysis. *Acta Cryst.* **A28**, 629–635.
Williams, D. E. (1973). Coulombic interactions in crystalline hydrocarbons. *Acta Cryst.* **A30**, 71–77.
Williams, D. E. & Cox, S. R. (1984). Nonbonded potentials for azahydrocarbons: the importance of the coulombic interactions. *Acta Cryst.* **B40**, 404–417.
Williams, D. E. & Houpt, D. J. (1986). Fluorine nonbonded potential parameters derived for crystalline perfluorcarbons. *Acta Cryst.* **B42**, 286–295.
Willis, B. T. M. & Pryor, A. W. (1975). *Thermal Vibrations in Crystallography.* Cambridge: Cambridge University Press.*
Winkler, B. & Buehrer, W. (1990). The lattice dynamics of andalusite: prediction and experiment. *Phys. Chem. Min.* **17**, 453–461.
Winkler, B., Dove, M. T. & Leslie, M. (1991a). Static lattice energy minimization and lattice dynamics calculations on aluminosilicate minerals. *Am. Min.* **76**, 313–331.
Winkler, B., Dove, M. T., Salje, E. K. H., Leslie, M. & Palosz, B. (1991b). Phonon stabilised polytypism in PbI_2: in situ Raman spectroscopy and transferable core–shell model calculations. *J. Phys: Cond. Matt.* **3**, 539–550.
Winkler, B. & Dove, M. T. (1992). Thermodynamic properties of $MgSiO_3$ perovskite derived from large scale molecular dynamics simulations. *Phys. Chem. Min.* **18**, 407–415.
Woodcock, L. V. (1971). Isothermal molecular dynamics calculations for liquid salts. *Chem. Phys. Lett.* **10**, 257–261.
Woodcock, L. V. (1972). Some quantitative aspects of ionic melt microstructure. *Proc. Roy. Soc.* **A328**, 83–95.
Woods, A. D. B., Brockhouse, B. N., Cowley, R. A. & Cochran, W. (1963). Lattice dynamics of alkali halide crystals. II. Experimental studies of KBr and NaI. *Phys. Rev.* **131**, 1025–1039.
Woods, A. D. B., Cochran, W. & Brockhouse, B. N. (1960). Lattice dynamics of alkali halide crystals. *Phys. Rev.* **119**, 980–999.
Worlen, F., Kalus, J., Schmelzer, U. & Eckold, G. (1988). Phonon dispersion in a deuterated 2,3-dimethylnaphthalene crystal at 123 K. *Mol. Cryst. Liq. Cryst.* **159**, 297–314.
Yamada, Y., Shirane, G. & Linz, A. (1969). Study of critical fluctuations in $BaTiO_3$ by neutron scattering. *Phys. Rev.* **177**, 848–857.

Index

Printed in the United States
By Bookmasters